The YOOOGE PHILOSOPHY OF LIFE

The YOOOGE PHILOSOPHY OF LIFE

By

Matthew John Corcoran

Published by Life Skills Australia

THE YOOOGE PHILOSOPHY OF LIFE
First published by Life Skills Australia. November 11, 2011.
(02) 8005 6777. International +61 2 8005 6777.
PO Box 30, Merimbula, New South Wales, 2548, Australia.
www.lifeskillsaustralia.com

Written by Matthew John Corcoran
Co-written by Universal Intelligence
Edited by Steven Shackel

All of the information and materials contained in this publication are for educational purposes only, and are not in any way intended to be used as a substitute for professional counselling, medical advice, spiritual guidance or any other such treatment, advice, or care. Always seek counselling from a professional counsellor, medical advice from a qualified medical or health practitioner, and spiritual advice from your own soulful-intuition.

Version 120622.

National Library of Australia Cataloguing-in-Publication entry: (pbk)
Corcoran, Matthew.
The yoooge philosophy of life / Matthew John Corcoran.
9780980645224 (pbk.)
Includes index.
Meaning (Philosophy)
Life.
Dewey Number: 121.68

Cover and text design by Life Skills Australia.
www.lifeskillsaustralia.com
Illustrations by Matthew John Corcoran.
Printed in Australia, UK and USA.
Distributed worldwide by Friends of Yoooge.

For those who *want* to know.

CONTENTS

FOREWORD

Matt Corcoran flashed into my life on TV. I didn't catch his name but he was one of the best standup comedians I had ever seen; world class. Great comedians tell more than jokes, they bare their souls in public. There are no successful, unintelligent, unempathetic or unobservant comedians. What I remember most about that TV appearance was that the comedy genius I had just watched announced that he'd sold up everything to "go on the road" as a journeyman Blues Musician. I saw him play eventually and found him to be one of the most entertaining blues performers I have seen, totally committed and a true professional. A pattern was emerging ...

A few years later Matt very kindly filled in for me at the Australian Blues Music Festival when I became ill. After playing till the early hours of the morning he turned up on my doorstep the following day to see how I was and asked if he could help me in any way before he left for his next gig. I already knew he was a gifted and very special man but this act of kindness really affected me. I felt I had known him all my life.

Other than music we discovered we had both studied and worked in graphic design and had much in common. More importantly, we had both concluded that most humans walk the planet with eyes closed to the myriad possibilities offered by life. Matt discovered I was an author and editor and mentioned he would need to take time to write down some of his experiences and thoughts about life, religion and spirituality – another shared and important part of our lives.

It didn't occur to me that this award winning musician at the peak of his career, with a statue in his honour and huge fan base to prove it, would become so impassioned by his spiritual experiences that he would walk away from the security of a burgeoning music career.

After eight years of writing he sent me a first draft of the first in his series of huge (Yoooge!) books. I know that an author's life can be difficult prior to successful publication so I offered my editing skills to assist with his book. It was no less an act of friendship than Matt would have offered. I am pleased to say he accepted my offer and am honoured to be first his friend and now his editor.

Matt is a force of nature. He writes as he speaks. He's amusing, kind, honest, irreverent and highly intelligent, as you will discover. His ideas and manner of presenting them will ruffle the feathers of those who are not open minded. Take a deep breath, expect to be surprised, even shocked and you will find you are on the ride of your life. Are you brave enough and open minded enough to accept new ideas and even timeless ideas offered from a completely different perspective? If you are, you will be rewarded with your own Yoooge Experience!

Steven Shackel, 2011

ACKNOWLEDGEMENTS

It's difficult to acknowledge those who helped me on my spiritual journey because *everything* in life is enlightening in some way. So who's responsible for what awakening in whom and why or when or how much? Who do you thank? Who gets the credit? For example, do I thank Eckhart Tolle for 'The Power of Now', do I thank the stranger who cut me off in traffic for the *power of now,* or do I thank my friends and family for the many moments of joy that truly *are* the *power of now?*

Spirituality is so subjective, yet objective, also elusive (and illusive) and mostly beyond our comprehension. It's like rice confetti at a wedding. Some sticks to your face, some lands in your hair and much of it misses you altogether because you were not in the right place at the right time. The latter ends up on the ground and is often eaten by small birds.

So I decided instead to acknowledge the people whose rice was so fiercely thrown into my face that it left dents in my skin.

The first person I would like to thank is Neale Donald Walsch. His book 'Conversations With God: An Uncommon Dialogue (Book 3)', inspired the Yoooge experience of November 2003, which then instigated an eight-year journey of spiritual discovery, most of which is documented in this book.

The second person is Esther Hicks. Esther translates (channels) a non-physical group-consciousness known as

Abraham, to a live audience. The audience asks Abraham questions and Esther translates the answers, which are truly amazing. The teachings of Abraham have not only changed my life but they helped clarify and validate the Yoooge Philosophy. It kept me sane.

I would like to thank the other fierce, rice-throwing authors by acknowledging them at the end of this book, in the 'Books That Changed My Life' reading list.

I would like to thank my naturopath and friend, Gary Hancock, for healing me. I went to see Gary many years ago as an angry young musician with a bad case of glandular fever. His naturopathic brilliance not only healed the glandular fever (in eight days!) but Gary's subsequent guidance, along with a hefty dosage of powerful Vega testing and homeopathy, saved my soul. He then slowly and ever-so-cleverly helped me rediscover my balance, my joy, my true self and my spiritual purpose.

I would like to thank Steven Shackel, for the editing, clarity, fact finding and passionate discussions about the science of communication. I would like to thank the incredible proof-reading super-team of Rosilyn Kinnersley, Ellany Whelan and Jo Jolley, who painstakingly proof-read and clarified my manuscript, word by word, many times.

I would like to thank my Mum and Dad for just being awesome.

I would like to thank my Soul and the Spiritual Collective for co-writing this book.

And finally, I would like to thank my beautiful partner Jayne. I obsessed over this project day and night for almost eight years, whilst Jayne languished further and further into the background. But she never complained. Her patient understanding and loving support, combined with her joy for life, kept me sane, balanced and positive. I could never have completed this book without her, and for that I am truly grateful.

Thank you.

INTRODUCTION

This book is like a total stranger coming up to you on the street and saying, 'Hey, did you know you're a gigantic doughnut?' You would probably think to yourself, 'Go away, crazy person' but the curious part of you hesitates. You suspend your judgment for a moment, stop fumbling about for the pepper spray and hear what the crazy person has to say.

He then goes on to explain that your digestive system starts with the mouth and ends with the anus and so your digestive tract is one gigantic continuous tube. Food travels into your mouth, down your throat, into the stomach, then into the intestine and is finally eliminated as waste. There are no gaps in the tube between mouth and anus, effectively making you a large body mass with a 'hole' in the middle! Therefore you *really are* a gigantic doughnut.

This book is a bit like that. It can be confronting and weird at first but if you suspend your judgment and allow the information to unfold, section by section, from front to back, then all will be revealed and it should all make sense. You may then discover that the 'crazy person' is not crazy but just trying to explain a very complex subject and needs time and space to do so.

You can still read this book with your can of pepper spray at the ready and if you believe this book is threatening your safety then hold those words face-up on the pavement and spray them in the eyes. Otherwise I recommend an open mind, since what I am about to describe would appear to a large proportion of the population to be 'out there, fruity, weird-arse ramblings of a crazy person'.

THE GOD WORD

I should point out early in the book that I use the 'God' word ... a lot! About four hundred times, the last time I counted. Don't panic. This is *not* a religious textbook *nor* am I affiliated with any religious group or spiritual organisation. I'm neither going to crap on about God nor bore you with spiritual-speak about the 'oneness' of the 'nowness' of the 'allthatisnesses ... ess'.

This book is simply my story inspired by a real out-of-body experience that took me to a place where dead people go. It explains the experience and the spirituality behind that experience.

Spirituality is not religion. Spirituality is an individual's search for higher truth that's personal, internal and doesn't involve any external group or organisation. You do it yourself from *within,* using your own mind. Religion and spiritual movements can certainly help you do that but you don't *need* a religion or a spiritual movement to do that. You have all the tools you need within your own mind to find higher truth and you can do so in the comfort of your own home, wearing tracksuit pants and sipping coffee.

As a writer I cannot discuss spirituality without using a God-type-word. I considered many alternatives such as 'The Creator', 'The Universe', 'The Oneness-nowness-all-that-is-ness-es ... ess' but I eventually settled on the word 'God'. It's simple, quick, and literarily convenient and it's actually quite a beautiful word if you remove the religious fanaticism connection.

But 'Oh God!' writing about God can be so confusing, especially when you consider what God *is*, what God is *not* and how that contradicts the common view of God. I couldn't just use the one single world. I had to use *three* different versions of the word:

- **'GOD'**: that which *is* the real GOD. GOD*like* refers to anything that is more *like* the true concept of GOD and not-GOD refers to that which is not *like* the true concept of GOD. It's also the *only* Divine element that gets a capital personalisation such as GOD but as well as 'Her', 'She', 'I', 'It', etc.
- **'god'**: that which people *think* is the real GOD, commonly known as 'God', the dude with a beard sitting on a cloud that's described in the Bible. It gets a lower case 'g' because *that* idea of god is essentially flawed (explained later) and so does not deserve a capital 'g' - in my book.
- **'God'**: refers to the mythical Gods, the more balanced concept of God/Goddess, the Soul (Higher-Self) and Soul-Group. They all get a capital 'G' because those concepts are much closer to what *should* be worshipped as a deity(s). As do other GODlike elements such as Mother Earth, the Moon and the Sun.

I suggest studying the above definitions for a moment; maybe have them tattooed onto your upper forearm, because they are vital for the understanding of the Yoooge Philosophy of Life.

As you may have already guessed, the GOD I am about to describe to you is not the male god, sitting on a cloud, with a beard, punishing Christians for thinking about group-sex. On the contrary, the GOD that I refer to in this book is vastly different and a whole lot nicer. She is mind-blowing and doesn't have a beard!

I also make no apology for the use of the words 'Her' and 'She' in relation to GOD. Although the true principle of GOD is neither male nor female but a perfect combination of the two, I am just doing my little bit to restore the balance of power in the male-female perception of divinity.

Let's face it, which one of the sexes do you think better represents the true nature of GOD? Women are naturally gentle, compassionate, nurturing beings. Men are good at lifting heavy things, drinking beer and stabbing meat - they tend to be more aggressive, less altruistic and somewhat emotionless!

Women are more *in touch* with themselves, whereas men are more likely to *touch themselves*.

I also choose the word 'She' because 'She' is actually a combination of the words 'he' and 'she' (s/he). If this still offends you then you really need to get over yourself.

I also prefer the word 'Cosmos' over 'Universe' since the universe, as in planets, stars and galaxies are only part of the Cosmos. The word 'Cosmos' is a Greek word and refers to the harmonious structure of the Cosmos of worlds within worlds.

IT'S ALL ABOUT YOU

A major part of any spiritual journey is to *know thyself*. Consequently, I am clearly and bluntly going to point the finger straight at 'Thyself', which is you. I use terms such as 'You think this …' and 'You do that …' and 'You are this …' not because I'm mean and superior but because it draws *you* into the book and makes it personal. Your spirituality is all about you, after all!

This defies many of the counselling recommendations in the 'Oh So Great and Wonderful Book of Counselling Recommendations'. It's considered rude to 'you' a client by implying that, 'You do this' and 'You do that'. I can imagine counsellors all over the world misting up their glasses as they hiss 'Hey! He used the word '*you'! …* hsssh!'

The more polite way to counsel is to ask you, 'How do you feel about a particular situation?' This gets you to become emotionally involved in your own dramas so that you self-discover your own faults. But screw that! Who has the time? I prefer the no bullshit, straight down the line, 'You do this because you're a flawed human-being like the rest of us' approach. It's the 'You need to wake up to *yourself* in order to *know thyself*' approach to spiritual literacy.

IT'S YOOOGE?

Something absolutely amazing happened to me in November 2003 and it changed my life forever. It was yoooge! That's actually where the word 'yoooge' comes from. Yes, there is no more spiritual significance to the word 'yoooge' than it's my Australian over-pronunciation of the word 'huge', pronounced 'yuuuge'.

Yoooge means big. How big? Very big. As big as GOD. As big as the Meaning of Life. As big as the interconnected, multi-dimensional infinity of the Cosmos. Throughout the book I identify many things that are yoooge to help define the bigness of yoooge by saying 'Now, that's yoooge!'

MOUTH FULL OF DRY BISCUITS

Yoooge falls under the banner of metaphysics. 'Metaphysics', according to the Oxford Dictionary, is 'The branch of philosophy that deals with the first principles of things, including abstract concepts such as being, knowing, identity, time and space … abstract theory with no basis in reality'.

No person, no book and certainly no group are *ever* going to be able to explain *that* fully! Trying to understand let alone explain the infinite multidimensional complexity of the Cosmos is like trying to play harmonica with a mouth full of dry biscuits. Sure it's possible to get a few biscuity type notes but it's mostly 'pssh, psshik, pssh, haarssh, pssh, psshik, pssh … f$#@!'.

This I know as being the only true fact. The Cosmos is, and the concept of GOD is, just too yoooge! Consequently, this book is a deeply spiritual, non-religious, slightly fruity, metaphysical *theory* about creation, life and death. It's a *theory* about the other cause of things. It's a *theory* to incite debate and get people talking. It's a *theory* to encourage the mind to question reality. It's a *theory* that tries to describe the indescribable, explain the unexplainable, all of which is pretty much unbelievable.

It is not literal and not to be taken literally and GOD help you if you *do* take this stuff literally. Have a look at the madness surrounding the people who take the Bible literally.

I would love you to joyfully contemplate what I have written, as if we're having a deeply spiritual conversation at a dinner party, albeit a seventeen-hour long conversation where I do all the talking. I don't want you to take it *that seriously* that you stand up, throw your napkin down on the floor and then climb *onto* the dinner table and kick over the cheese fondue - all the while spitting and mumbling something about irreverent literacy and cheese on your boot.

On one hand, the three most powerful words a human can say are 'I don't know' since it keeps you open to *all* points of view. *On the other hand,* I saw things in November 2003 that I might never write about because it's just too unbelievable. *On all the fingers of the hands* (the details), I have researched and pondered this stuff for years, added my own theories and combined theories from other authors.

But I can't prove any of it! I had a yoooge experience that changed my life forever but I didn't get a receipt. I didn't manage to sneak in a couple of digital photos of me hanging out in the spirit-world. I didn't manage to grab some of the Cosmos, stuff it in a jar and then post it back home hoping customs wouldn't open the lid and spray it with pesticide. I didn't get a signed poster saying, 'To Matt, best wishes, GOD'.

I can't even describe what I saw fully because what I saw was viewed from outside the human perspective and that includes being outside the limits of the English language. There just aren't enough English words to describe what I saw and so I made a lot of them up, which you'll discover as you read on. There is also a glossary of such words and terms at the end of this book.

And even if I could explain it fully with lots of gargantuan words, like 'gargantuan', it would still be ridiculous to assume that the Cosmos is either *this, or it's that*. Definition is a human trait. The Cosmos is a pulsing, conscious *illusion.* It's alive, it thinks for itself and it changes to suit whatever form, function and phenomenon is required, including my gargantuan definition, your gargantuan definition and anyone else's gargantuan definition. In that regard, everything is true!

The Cosmos is what's known as a counterintuitive Divine dichotomy - it is this, yet it is that, yet it is neither, yet it is both. Any good description of the Cosmos *should* be awash with contradictions and rational-irrational conflicts. I think I have mastered the contradictions and rational-irrational conflicts part.

WHY BOTHER?

So why bother trying to explain that which is unexplainable? The real truth is, I bothered because I couldn't help it. Several months after November 2003 I had this cute idea that maybe I should write a little ol' book about my experience. Shortly after, that idea turned into an all-consuming passion that literally took over my life. I became obsessed like those people building miniature mountains in the film *Close Encounters of the Third Kind,* or Ray Kinsella (Kevin Costner) in the movie *Field of Dreams* - 'If you build it, he will come'.

I feverishly wrote from early morning to late at night, almost every day of the week, between part-time jobs that helped pay the bills, as I built my literary 'Noah's Ark'. My hands often had trouble keeping up with the information that came to me. My beloved partner languished farther and farther into the background, living within the madness of spiritual writing.

Every time I thought I'd finished, I would get another 'download' (spiritual download) of information. I'd then add this new clarity to the book, which effectively meant rewriting the entire book. I have since re-written this book twenty-something times, to the point where I now know every paragraph intimately.

The book ended up being over one thousand pages long so I recently split that book into three books called 'Yooоge Book One', 'Yoooge Book Two' and 'Yoooge Book Three'. The three books combine into a set called 'The Yoooge Project'. This book is 'Yoooge Book One'. Yoooge Book Two and Three will be released in the near future. For more details, see the back of this book.

IT DOESN'T MATTER

Some of the chapters in this book are a bit 'chewy'. I have tried to explain them as simply as possible but the Cosmos is complex. I couldn't avoid some mind-boggling metaphysical and scientific content. There are only a few chapters like this. The rest of the book is tender and delicious. I would still suggest reading these chapters slowly and 'chewing' on the information thoroughly to avoid mental indigestion later on. But don't worry if you don't understand it. Heck, I wrote it and I'm still trying to understand it.

So how do I know I am right? Well I don't. More precisely, I know it's right *for me* but I don't know if it's right *for you.* You have to use your own intuition to decide whether my irreverent theory is right for you, or nonsensical whoopy foof.

Then how do I know it's right *for me*? I often answer this question by asking people, 'Do you love your kids?' Most people say, 'Yes of course I do'. So I say, 'Well prove it to me, right here and now'. Most people reply with, 'I don't have to prove that to you. I just *know* I love my kids'. In much the same way, I have this knowing about the 'The Yoooge Project'. I just *know* what I have written is right ... *for me.*

In the bigger scheme of things it doesn't really matter if I'm right or wrong, it doesn't really matter whether I have explained myself properly of whether you understand it. It doesn't really matter if you think this book is nonsensical whoopy foofness.

Why? Because the wonder and enlightenment of what I saw in November 2003 has been energetically encoded into the atomic structure of this book. Within the paper pages of the printed edition and within the binary code of the eBook, lies a buzzing electromagnetic energy that holds the fundamental Truth of what I am trying to describe. If you're open to this information, then that energy will resonate with your own energetic framework and become *part of you.* You will then have your own unique yoooge experience, at whatever level is *perfect* for you.

I can appreciate how this might sound like a rather big claim to make. Like, 'I'm Super Spiritual Man' with powers worthy of my own TV show. But not really; I'm not doing the amazing bit! My intent is to give you, the reader, a yoooge experience. That's my role. That's all I have to do. The 'amazing bit' is done by my Soul (Higher-Self). *It has* the creative power required to encode the energy of yoooge into this book and then conspire with *your* Soul to give you a yoooge experience. It can and it does this easily since it *is* 'Super Spiritual Man'. So is *your* Soul. So is *everybody's* Soul.

I also know that your Soul and my Soul hang out together in the spirit-world. They have already conspired together to get you to read this book and they've already conspired together to organise what happens to you, *after* you read this book.

**Metaphysics is a theory
that doesn't have to be proven …
but it does have to be beautiful.**

THE BLUES STORY

My story starts with music. I used to be a Blues, slide-guitar player and I toured Australia full-time for many years, playing four or five shows a week, both as a solo artist and with a band. I had a record deal, three successful albums and the sort of lifestyle that people dream about - fame, fortune and free beer. But that's just what people dream about.

In reality, being a touring Blues musician is bloody hard work. It's long hours, low pay and thousands of kilometres of driving in an overloaded, stuffy, van/truck, complete with overweight bass player farting some sort of rotting duodenum death spray.

I did well as a musician but I wasn't *that* famous, or at least not the sort of fame people associate with musicians. I was a medium sized fish in a small niche market. As far as record sales are concerned, my claim to fame was being Number 1 in the Blues Charts for ten weeks. That's the *online* Blues Charts. The online *Polish* Blues Charts. The *independent* online Polish Blues Charts.

I certainly wasn't rich. I sucked up to famous people just to look more famous-er - many of whom brushed me off as a desperate try-hard trying to be more famous-er. I didn't take drugs. I didn't drink beer. I didn't like pubs. I hated all the attention and I didn't like loud music. Sort of makes me wonder, 'What the hell was I thinking?'

The beer was indeed free but I didn't drink beer. I also had a couple of sponsorship deals and so my guitars and amps were also free, but I had to struggle like a single mum on a pension supporting five kids, to secure those deals.

As for the groupies, my gigs were mostly filled with drunk, middle-aged men. That squashed the groupie 'dream' like a four-door sedan dropped on a sandcastle of icing sugar! Don't get me wrong, I loved those guys, just not in *that kind of way!* They were my life-blood and they were passionate Blues fans, many of whom became my friends. But they were men. They were my man-groupies. Not the sort of gorgeous, screaming, fanatical girl-fans you associate with Pop bands.

My man-groupies were always drunk. Everyone was drunk, except me because I didn't drink. Drunken people can be very annoying when you don't drink. I even tried getting drunk, for a while and then the drunken people became hilarious. But I couldn't maintain this drunkenness four to five nights a week, so I stopped drinking and the drunken people became really annoying again.

They yell passionately (pansshnootlateeey) about all sorts of stuff and their spittle lands in your drink/face/hair. Their friendly back-slaps are more like Heimlich manoeuvres. They stagger about wrecking expensive music equipment. Then they proceed to beat the %$#@ out of each other.

Blues bars around the world are well known for their bar fights. Outback and country Australian Blues bars are no different, if not a little bit wilder. For example, I once had to stop my show to wipe blood off my face from two people brawling. Fair enough, that happens, but these two people were women! No offence to women but jeez, you normally don't expect a woman to king hit another woman in the face because she stole her spot on the pool table.

The woman who was hit went down hard, splattering blood across my music gear and my face. The other woman just continued to watch my Blues show, bopping from side to side, smiling, as if nothing had happened. Worse still, everyone around her just continued to watch my Blues show, bopping from side to side, smiling, as if nothing had happened. That's when you know you're in a *bad* pub - where face-punching girl violence is considered 'normal'.

I stopped playing long enough to wipe potential hepatitis molecules out of my eyeballs but that started *another brawl*! So I thought it best I keep on playing. I had blood on my face all night until I got back into the motel and showered in bleach and Holy Water.

The bad fights usually continued outside as street fights. Often, my gigs were hosted in a two or three-storey pub. One of my more 'sick' although 'enjoyable' pastimes was to race upstairs, stand on the balcony and in relative peace and obscurity watch the humans' battle. It was like boxing, only drunk with little fighting skill, lots of on-the-ground wrestling and a fair amount of yelling.

From my safe balcony viewpoint, I've seen just about every type of negative behaviour. I've seen people's faces smashed through windows. I've seen couples vomit on the ground, pause, wipe and then continue to tongue-kiss each other. I've seen all sorts of illegal drug use, handguns tucked into waistbands and heard every version of the words %$#@ and %$#@ and you %$#@en %$#@. I've seen drunken stunts; drunken arguments and drunks run over and killed.

I once witnessed a guy (who was so blind-drunk he could barely talk) stand up, fall over, stagger sideways and *then* hop into a fully worked 500 horsepower V8 car and drive away, sideways, no lights, at night, through the fog at 200km/h. GOD knows how he got home but he did!

I WAS ANGRY

When you sign up to be a Blues musician you sort of expect beer, brawls, bikies and burnouts. It's on the application form, 'Tick here to accept the drunken terms and conditions …'. It was the other stuff that wore me down.

I was 4000 kilometres from family and I missed them. I missed my nephews and nieces growing up. I missed my good friends and most of all I missed having a normal, healthy relationship. Blues music, touring and girlfriends have historically never combined well. I missed having a home base. I missed having a normal life. Shit … I missed just having a drawer to put my socks in. I missed *normal*.

I was constantly immersed in every possible version of 'pub smell'. That aromatic combination of alcohol, cigarette smoke, ashtrays, mouldy carpet and the tacky perfume of the industrial cleaner used to clean the vomit, ashtrays and mouldy carpet. My budget only allowed me to stay in cheap motels, which meant I breathed in people's dead skin cells, mouldy air-conditioning fumes and more of that tacky perfume of the industrial cleaner used to clean the ashtrays, toilet and mouldy carpet (and on a bad day, vomit).

I tackled mosquitoes, flies, bedbugs, spiders, the occasional snake and every other 'I'm going to bite your head off' critter that lives in rural Australia. I would go to bed every night at about 2-3 a.m., on a full-stomach after eating re-heated pub style food, since the only time you get to eat is after the show. I was woken *every* morning at about 5-6 a.m. by old people (why do old people get up so early?) standing outside their motel rooms, deafly *yelling* gossip at each other, such as 'Then Doris said … and then *he* said … then *she* said … ' answered by a chorus of, 'Oh I knoooooow', which sounds remarkably like a barn full of chickens.

I'd go back to sleep, fuming, only to be woken again by the house-keeping, cleaning staff barging into my motel room at 7-8 a.m., *tripping over* the 'Do Not Disturb' sign on the way in, to see my bare naked butt lying on the bed. I would go back to sleep, fuming. In hindsight I think the house-keeping ladies may have been the chickens. Even to this day, I have trouble sleeping without a pair of boxer shorts on, in case 'they' come barging in.

I was physically and emotionally exhausted and constantly frustrated. My diet was pretty ordinary and so my body was becoming increasingly unhealthy. I was lonely, bored and depressed. I was constantly immersed in the dark, negative energy of pubs and clubs, which osmotically infiltrates the semi-permeable membrane of your 'niceness energy field' and becomes part of you!

The outcome of all this was it made me angry … very, very angry. My body was big (one hundred and ten kilograms) and angry. My shaved head and tattooed arms made me *look* angry. My attitude was '%$#@ off and leave me alone, I'm angry'. There was so much anger within me that I *scared* people - other big, angry, scary people. I remember one night when I was hiding in the toilet (only place in a busy pub to get some privacy when your face is on the poster) and I heard this big guy come in. I recognized his voice, as I had met him earlier. His name was 'Spike' or 'Smasher' or some other 'I will hurt you in a hundred ways' kind of nickname.

I heard him say to the guy next to him at the urinal, 'How ya goin' mate (Pause)? Jeez, that Matt Corcoran's a mean looking bugger. I wouldn't want to pick a fight with him. Yok. Yok. Yok'. 'Shit!', I thought to myself, 'These guys *scare me* and yet this one hundred and forty kilogram man-mountain, with arms like Christmas leg hams was worried about *me,* hurting him. What kind of person have I become? This isn't me. I'm nice. I'm the happy guy!'

But I wasn't. I had become 'An Angry Young Man', which also happened to be the title of my second album. I remember the day when I decided that I needed help with my anger. I was in Melbourne, in a pub, setting up my music equipment before a show. It was about 4 p.m. and some drunken guy decided he was going to ask me a whole bunch of stupid smart-arse questions. That poor guy was in the wrong place at the wrong time and copped about thirty-three years of pent up aggression. I snapped, turned and pushed him with so much force he actually back somersaulted over three tables, broke several glasses, a chair and part of his body. Suffice to say he stopped asking questions.

I was practically oozing negativity. The worst part was, I had such little self-awareness and was so busy trying to be famous that I had no idea. I was being a typical Aries – flat out, head down, racing at ramming speed towards the goal, knocking people out of the way to get there (literally), never stopping, never resting. Yet all I was doing was moving farther and farther away from *my true self*. My life was going *backwards*. Then fate stepped in.

THE AWAKENING

On 25 October 2003, I competed in the finals of a Blues music competition where the winner would represent Australia in the International Blues Challenge (IBC), held annually in Memphis, USA. Each musician or band performs in front of a panel of judges, which I'd already done and consequently made it through to the finals.

If I won the Australian final, I would be flown to Memphis, USA, all expenses paid, to compete at an international level in the IBC itself. This was a massive opportunity for me to break into the USA Blues scene, the *home* of the Blues. I thought I was a 'shoo-in' to win the Australian heat. I believed it. My record company believed it. Most of my fans believed it. My management was praying for it.

I thought I had it all sorted out - win the Australian IBC final, fly to America, win the USA IBC, conquer the USA Blues scene, then get heaps more fame, fortune, free beer and spitty man-groupies. Then I would be 'set for life'.

On the night of the Australian IBC final I played the best set of Blues slide guitar I can ever remember playing. I thought to myself, 'Hey buddy, that's it. You've done it. You've won'. My friends in the crowd thought I'd won. My record company thought I'd won. My management was buying strangers drinks.

In my head, I was packing my bags to go to the USA. But ... (pause for effect) ... I didn't win. I didn't even come second. I came third! Blerrbrrpghpapapapasshblah. That was the sound of my ego deflating.

Shortly after they announced my name in third place, I heard those crappy statements people often make to try and cheer you up, such as 'Wow, I was *sure* you were going to win!' or 'Oh well, you've still got your health!' and my favourite, 'Hey, don't worry, next year you might get second place and then the year after you might get first place!' I didn't show my anger. I just rammed that anger down deep inside, with all the other anger. Grrr!

The statement that really knocked my ego onto the cold hard floor, quickly snuck around behind it, placed it in a double-arm choke-hold and then squeezed it until it went red in the face and passed out, was the record company boss saying, 'Hey. Bad luck. You didn't win. That's OK. Just hang in there mate, another couple of years and then you'll be ready to start gigging (performing) in the USA.'

'What!' I thought to myself. 'Just another couple of years and *then* start gigging in the USA? I'd *already* been doing this for ten years and (singing) 'It's beginning to look, a lot, like, shit-mas ... aaalll aroouund my worrrld!' That realisation was enough to make me stop, lift my Aries-ram head, look around and then ask that all important question, 'What the %$#@ am I doing with my life?'

THE STRAW

In the last few years of my music career, I toured in a three tonne truck, with a bed and kitchen in the back. I called it 'Zena the Kangaroo Warrior' mainly because I had decked it out with a massive cast iron Roo-Bar (Kangaroo Bull Bar) so I could run over the occasional kangaroo, wombat, camel, spider, without damaging my truck, or me.

I was driving Zena away from the IBC fiasco, ego passed out on the passenger seat next to me, when I suddenly had an 'Oh %$#@!' moment. Everybody has an 'Oh %$#@!' moment - a sudden shocking realisation that a major part of your life is not working out for you and you're going to have to do something about it. My 'Oh %$#@!' moment was realising that I didn't enjoy my music career anymore.

Music was my passion. It was my greatest joy. I would practise slide guitar for hours on end until the day just seemed to disappear ... because I loved it so much. I got sacked from my first design job because I kept taking time off work to practise ... because I loved it so much. I came second in the '1994 Marshal Amps, Guitar Solo Championships, Perth, Western Australia', after *only ten months of playing* ... because I loved it so much.

But now the joy had gone, only to be replaced with the deadlines, expectations and turnover of the music *industry*. I was frustrated, I was bored and I was miserable. I realised that the only reason I was still gigging was to keep everybody else happy - mainly my fans and the record company. They supplied the approval and I was addicted to it. I had become a 'clap-hands', 'Whoohoo', junkie. To climb the greasy pole of fame, I had traded my joyfulness for some imaginary expectation of fame and fortune that I didn't even really want.

The gloss of being a famous Blues musician had suddenly been sandpapered clean off my goals by a sudden and hard 80-grit awakening. It was like one of those moments in life that seems like a great idea at the time but turns out to be a sad tragedy. Like that time you hired an inflatable bouncing castle for your son's fifth birthday party but it accidentally deflates, temporarily entombing fifteen five-year-old kids inside a dark, claustrophobic, plastic prison. Now, anytime birthday cake is mentioned, the neighbourhood kids quiver in fright and mumble, 'Not the castle, mummy, the walls … '.

My music-bouncy-castle started out well, my intent was good, I was excited but it very quickly turned to shit. My grand idea deflated and entombed me in negativity. Just like the father of the bouncy-castle-kid, I couldn't help but feel a bit regretful, disappointed and sad about my idea of becoming a musician. Now, anytime someone mentions the idea of going back on tour, I quiver in fright and mumble, 'Not the pubs, Johnny (record company boss), the drunks … '.

So, I decided that *maybe* I should *think* about the *possibility* of *perhaps* getting out of the music business! I was strangely excited by this thought, almost joyful.

Then my ever-chatty, chaotic, stressed-out mind came barging into the room yelling, 'Hey, just what the hell do you think you're doing? You can't just leave music. Think about all the hard work you have done. Think about how far you have come. Think about the fame, the free beer and the spitty man-groupies. What about John (John Durr is head of the Record Company, Black Market Music)? He's been good to you (he had) and become a good friend (he was). What about Boyd (Boyd Kelly Management) and all the gigs he's already booked? How can you let them all down?' So I decided to stay a little bit longer. This thought made me feel depressed, stressed and flat.

'Phisst, phisst, phisst' - the sound of me trying to manually inflate my deflating music-bouncy-castle.

A week later, Wednesday afternoon, 29 October 2003, I was reversing Zena, the little Blues truck, into a parking bay at the motel attached to the pub that I was performing in later that night. I was still thinking about getting out of the music business ... or maybe not ... or soon ... or maybe after ten years ....or ... (Crash!) I'd reversed my truck into the roof of the motel, smashing the top right-hand side tail-light that I had only just replaced a few days before.

This was no big deal. My truck had lights all over it and I'd broken many of them before. But as I was picking up the little pieces of broken red tail-light-plastic from the ground, getting more and more pissed off, I thought to myself, 'That's it. I've had enough. I'm sick of all this crap! I'm leaving the music business for good!' Even today it strikes me as weird how *that* was 'the straw that broke the camel's back'.

A few minutes later, I sat on the motel bed pondering life. I thought to myself, 'If I am going to leave the music business, what will I do? What will be my next adventure?' Then a thought suddenly popped into my head, 'Maybe I could become a healer?'

THE IMPASSIONED PLEA

Prior to this day I was very much into the whole new-age, spiritual, alternative 'thing' or 'NASA' for short. As I travelled around as a musician I made it my mission to meet and talk with as many NASA people as I could. I also read many of the more popular NASA books, had dozens of psychic readings/healings done, saw a ghost or two, did Reiki, and even had a go at being a medium for a few months in the online chat rooms.

But I was most fascinated by the idea of healing. I wanted to heal people. I wanted to take away the pain I had seen displayed by the humans at my shows, as they punched, spewed and yelled at each other. I wanted to make them *better*.

The entire ten years I was a musician I observed human behaviour, almost as a hobby, and not just the drunken, spitty, fighty ones but all human behaviour, including my own. I became fascinated with why humans behave the way they do, how they cover up their pain and how their negativity affects them.

I felt their pain - literally. I had already earned a Bachelor of 'Get %$#@ed' from the University of Life Experience, majoring in anger, frustration and sad-rage. Being a musician with some fame, I had also obtained an Advanced Diploma in Ego-centric Behaviour, specialising in emotional-pain-suppression and how the wounded-inner-child part of us craves approval, recognition and friendship. I was also duly qualified in the fields of loneliness, depression and re-heating pub style meals.

I was ready so I did something, for want of a better phrase, spiritually intense! Some might say stupid, others might say risky but I like the phrase, 'spiritually intense'. That day, after I had pondered life for about an hour, I lay down on the motel bed, closed my eyes, cleared my mind (as best I could), took three big deep breaths and then I made the most impassioned, heartfelt plea to GOD that I can ever remember making in my entire life. I said the following words: 'GOD, I ask you to use me as a tool for healing'.

I had read that phrase somewhere, possibly Louise Hay or similar, and it felt good so I said it. But I didn't just say it. In those few words I'd committed myself to GOD to be of service to the world as a healer and I meant it ... *I really meant it*. There was *so much intent* behind those words that I can remember being just a little bit frightened about how it was going to pan out!

That's why it was 'spiritually intense'. You cannot combine that much *intent* with such a *powerful desire* without creating some mind-blowing outcome in your life. It's basic cause and effect. I didn't know that, nor did I think it through, especially the outcome. I was like a child with a box of matches, in a cardboard box storage facility, alone, bored and full of sugar! 'Whoohoo, matcheeezzz (scratch, scratch ... whoof!)'.

My idea of being a healer, or at least the one that stroked my luscious musician ego at the time, was of becoming 'The Anointed One' whose amazing hands-on healing powers would tantalise and amaze the masses as they sprinkled rose petals at my feet while chanting my name. Well not exactly, but you get the idea.

But this was not to be the case. Whatever Cosmic force received my impassioned, heartfelt plea had a different idea of what being a healer meant. 'Their' idea of healing was to write books.

Books! I had no idea at the time that my imaginary career as the 'Anointed One', complete with tunic, gold cross and TV show, would be replaced with the more humbling and challenging career as a writer. Many years later I realised why. I primarily write books that help people experience more joy and joy *is* the greatest healer.

I QUIT

The next day I rang the record company and told them I was quitting the music business for good (much to John Durr's dismay). I rang my family and told them I was quitting the music business for good (much to my mum's delight). I rang some friends and told them the same. Most of them were musicians and reacted with the sort of shocked disbelief you would reserve if a friend told you they were a Korean spy. 'You're what … really … why?!'

I couldn't quit immediately like I wanted to, because my very efficient management team had already booked me solid until March 31, 2004. So I decided to stay on and finish those gigs as a way to honour my fans, my gigs and my management. I decided that March 31, 2004 would be my last show – five months away! Meanwhile, I kept a close eye on shop windows looking for a tunic, gold cross and rose petal supplies.

Those last five months of touring were the hardest of my entire music career. I really hated it! I was so tired, bored, depressed and angry that every day was a struggle. Towards the end, I was so fed up with being a musician that I'd finish each show, run off stage and hide so the drunken people couldn't find me.

I remember one show in a crappy, dark, dangerous, inner city pub, where I didn't even finish the last song and just walked off stage, disappeared through a fire exit door and hid under a stairwell. I thought, 'There's no way the drunken people will find me here'. Little did I know that the stairwell was backed by tinted glass. As the crowd exited the pub they could clearly see me crouched up against the glass and so they banged on the window to get my attention. I could have cried but men don't cry. We just ram the sadness deep down inside so that it becomes sad-rage, powerful enough to entice you to push a man so that he somersaults backwards over three tables.

THE SIGN

After hiding under a stairwell you'd assume I would think, 'Jeez, you know what? I don't think this music-career-thingy is really working out for me, especially when you consider that I'm now HIDING UNDER A STAIRWELL!' Yet my ego was addicted to the approval I got as a musician. To leave music would mean the death of that ego (or at least part of it) and so my ego was struggling for its very survival. I fought my decision to leave music by urging my mind to race with indecision, 'Should I stay?', 'Should I go?', 'Have I made the right decision?', 'Am I kidding myself?'

My indecision mantra of 'quit, stay, quit, stay, quit, stay!' made me sick (I lost my singing voice in the last two months of that tour) and it almost drove me mad! Even up to one and half years *after* I left music my ego was still questioning my decision to leave.

Eventually, out of sheer exasperation, I handed it over to the Cosmos by saying 'Hey, GOD, I don't know whether I'm doing the right thing by quitting the music business and abandoning my music career. Can you *please* give me some kind of confirmation that my decision is the right decision for me? And I don't want some wimpy little 'coincidence'. I want a big, fat, black and white, knock you down, can't miss it, kind of sign'.

On 31 March 2004, my last show was cancelled due to the gangland shooting of Lewis Moran at the Brunswick Club in Melbourne, Australia. The Blues show I was meant to do was right next door and the entire area was cordoned off by the police as a murder scene. As no one could get in or out of Brunswick, no one could attend my show and so it was cancelled.

This was my big, fat, black and white, knock you down, can't miss it kind of sign from the Cosmos that my music career was ... DEAD, FINISHED, OVER! Pretty obvious really.

A few weeks later, I moved to coastal New South Wales, sold all my music equipment, my guitars, Zena - the lot. I chose absolute and complete disconnection. I never returned to the music scene and I have not played a note on my guitar since.

NOVEMBER 2003

Rewind back to November 2003. It's a couple of weeks after my impassioned, heartfelt plea in the motel room. I've gone back to 'work' as a musician to honour the last five months of pre-booked Blues shows and I'm now in Melbourne.

I had two weeks of gigs booked, mostly on the weekends. During that same period, friends of mine were travelling to the USA to record a Blues album. They had a house just south of Melbourne, on the bay in Dromana. It was beautiful, it was peaceful and I needed a rest. I jumped at the chance to house-sit for them, whilst they were away.

During that house-sit, some day between November 2 and November 13 (most likely November 11, 2003), I had an unexpected out-of-body experience that changed my life forever.

I was sitting on their couch reading. It was quite an enlightening book by Neale Donald Walsch called 'Conversations With God: An Uncommon Dialogue (Book 3)'. When reading such books I often get 'wow' moments. These are those moments when you read something so amazing that you just have to put the book down and go, 'Wow'. On this particular day I had a rather *big* 'Wow'. Walsch's words triggered some memory in my mind and I had a *massive* realisation, and then another and another. Suddenly a sort of portal opened up and whoosh! My conscious mind unexpectedly left my body, left the room, left this world and took off.

The room disappeared. I went up and into a funnel of light and was suddenly *surrounded* by light. I was standing in light, looking at light and feeling light. The light was everywhere and *everything*. I had one of those moments that you often hear car crash victims talk about, where time slows down and everything seems to be moving in slow motion. Except in my slow-motion-moment I remember saying something quite lame such as, 'Wow, everything *really is* made of light … wow!'

I immediately knew this was the light that religions refer to as 'the Light of GOD'. Light gets a capital 'L' out of Divine respect and to differentiate it from normal visible light. Visible light is a very poor comparison to this God-Light. Even sunlight is only vaguely similar. God-Light is atomic, alive, beautiful and yoooge! Imagine holding the power of a nuclear bomb between your hands … as Light. Imagine the answers to every single question ever asked as one clear thought ... as Light. Imagine all the life force and dynamic energy of all living things … as Light. Imagine every loving feeling ever experienced by every living entity from all time … as Light.

The Light was filled with power, wisdom and joy. It was fat with it, practically overflowing with it, almost as if the Light had form. It was as if you could drink it or mould it, though it was formless.

Even though I was surrounded by Light and there was *only* Light, I could still '*see*' stuff. The 'seeing' was like daydreaming. You know those times when you're sitting in a park, daydreaming? You might be blankly staring at a tree or the grass yet a daydream can manifest in front of your eyes in full detail as if the park isn't even there. You get lost in the daydream until something disturbs you and you suddenly remember where you are - sitting in the park. This was very similar.

I was totally mesmerised by this Light. I stared at this Light in the same way children stare at the big cats in the zoo - mesmerised, gleeful, wanting to touch the big furry kitty. So I 'touched the big furry kitty' *with my mind* and as softly as a child's hand touches something powerful, amazing and beautiful. My mind melded with the Light for a millionth of a millionth of a millionth of a second and then Boom! Wacko! Fasssh! Harrngghshhsh!

In that moment I became *one* with the Light and for a split second I knew everything you could possibly know about anything! I had a realisation the likes of which I have not had since and several years later I am still trying to understand. It was Yoooge.

Then, whoosh, I came back. I came back through the portal, back to Earth, back to the room and back onto the couch. My body was still there when I came back to the room, which was really handy! It never left. My mind had been on the journey into the Light and back again. As it slammed back into my physical body it created the same kind of experience as when you awake suddenly from a dream. Men know this phenomenon well, when their wife suddenly jerks herself awake and accidentally kicks them in the nuts!

HARRNGGHSHHSH

My entire yoooge experience was no more than a nonillionth of a second (1, 000, 000, 000, 000, 000, 000, 000, 000, 000, 000th of a second). One nonillionth of a second I was a 'child' patting the 'big, pretty, kitty' made of Light, then the next nonillionth of a second I was back in the room sitting on the couch, book in hand, with what I'm sure would have been a 'mad cow' look in my eyes. I was completely startled, traumatised and flabbergasted.

I'd read so much about other people's tales of enlightenment and spiritual visions and theirs always seemed to be more fun. They described Angels, celestial choirs and feelings of bliss. But I didn't get any of that 'lovey-dovey' good stuff. Instead, I got a massive dose of wisdom but without the love. I got the 'book' without the 'hug' and it shocked the shit out of me - so much so that I made a funny sound, 'Harrngghshhsh'.

The best way I can further describe 'Harrngghshhsh' is with emotion. Emotion is far more descriptive than any number of words. If you can *feel* it, then you can *be* it. Start by imagining the feeling of regaining consciousness. Those of you who have been knocked unconscious will be familiar with that sickening feeling, the moment you regain consciousness and suddenly realise *where you are*. If you've never been knocked out then maybe you should take up a serious contact sport or join a Blues band.

Imagine then, the sickening feeling you would experience if you were reading this book, blacked out and then woke up lying on your back in the middle of a soccer field. You're twelve years old; you've been knocked unconscious in a soccer game. There is a group of concerned adults standing above you asking, 'Are you OK?' Your head is buzzing and you suddenly realise that you *really are* a twelve year old child and you *really are* lying on a soccer field.

Your life so far, the life that you thought was real, including the part of your life where you were reading this book, *is not your real life.* That life has merely been the imaginary result from a few seconds within the mind of an unconscious twelve year old, lying on a soccer field. Your real life *is* the twelve year old boy. That would sum up most of the 'Harrngghshhsh' of the shock feeling. But it was more complex than that.

Secondly, imagine how you would feel after exploding with diarrhoea in a public toilet then realizing there is no toilet paper … and you're on your own … and your wearing white pants …. on the way to a wedding … your wedding! Now that's desperate panic.

Thirdly, imagine how you would feel if you were five years old and your parents dropped you off at a sports field to train. As they drove away you realise it's the *wrong sports field.* Your team is not there, and your parents won't be back to pick you up for hours. Now that's desperate childhood loneliness.

Just to confuse matters, imagine how you would feel if you returned home after a twenty year overseas travelling adventure and everyone you loved was there to meet you. They hug you tight and welcome you home. This feeling was the most surprising addition (for me anyway) - a weird homecoming, longing, remembering feeling.

Finally, imagine how angry you would feel if you'd worked on a certain project for decades and it was the most important thing in your life. You sacrificed years of family time and recreational time to get this project finished and then you find out it's not important at all. In fact, it's worthless. This is a weird, pissed-off, kind of depressing feeling.

Combine these five feelings together and multiply them until you feel physically sick. That's close to the feeling of 'Harrngghshhsh'.

Why was November 2003 so shocking? Why did I go 'Harrngghshhsh'? Why? Because of what I saw and what I came to realise in that nonillionth of a second. What was that? I can't tell you. Not in a few words anyway. That's the purpose of this book and it's going to take 389 pages to do so.

THE DOWNLOADS

The November 2003 experience didn't end there. Shortly afterwards, I started to get what I call 'downloads'. Downloads are large chunks of thought that *somehow* enter your mind, from a spiritual source much higher than ourselves (Soul).

I borrowed the term 'download' from the Internet. A spiritual download is like an Internet download, where a lot of information arrives *all at once*. I like to think of them as *within*loads because the information comes from *within* each person's mind via their Soul, but the term 'download' just sounds better when you say it.

You can easily distinguish between downloads and your own thoughts. Your thinking mind thinks in a linear fashion, one thought at a time. Downloads come as large chunks of thought. You just can't think like that! Downloads also come with a feeling that's 'superior yet humble', in a powerful, wise and joyful way. This feeling combines with the thought to produce a **block of thought**

accompanied by a feeling to create a knowing. These are the 'Wow' moments I described earlier. It's like you just know stuff, it's mind-blowing, it feels amazing and you have to go, 'Wow!'

Depending on the information in the actual download, there is usually a brief moment of understanding or a 'Wow' moment and then the realisation quickly fades. It's like you just suddenly know stuff and then it fades away. That used to be very frustrating but I soon realised not to panic. The information is still there, it's just that it unravels over time, piece by piece, to suit each individual.

November 2003 was my first download and definitely the largest download I have ever had. I still get downloads almost every day, some big and some small, which could either be the yooogeness of November 2003 slowly unravelling or a bunch of altogether separate downloads – I'm not sure. I was still getting downloads about this information the day before this book was handed to the editor, so I know there's more to come.

Anyone can get downloads and *everyone* already does. Some people experience downloads in the form of inspiration or the courage to 'go get it'. Others can get a series of weird coincidences, synchronicities and opportunities that 'guide them' through life. Others get direct answers to questions during meditation. Others can experience enlightening moments of Grace. Whichever way, it's your Soul 'talking' to you and everyone's Soul does this all day, every day - you just have to listen.

There is no special training needed to get downloads. It's easy, it's effortless and it's as natural as breathing. The single most important thing to remember with downloads is you have to *really want to know* the answer and you have to be *really clear about your question*. Then let go, stay joyful and the answer will come in whatever manner best suits you.

The only 'catch' is that full access to downloads can only be obtained in a state of joy. Downloads are super-positive, high-frequency sources of knowledge. Super-positive, high-frequency sources of knowledge need super-positive, high-frequency *channels* of communication. Joy *opens* a super-positive, high-frequency *channel* of communication within the mind. If you can laugh then you can get downloads. The happier you are, the more downloads you'll get and the bigger they'll be.

Positive thinking creates such states of joy. Unfortunately, humans have such a bad, subconscious habit of negativity that it's actually easier for most people *not to think,* than to think positively. That's why activities such as day dreaming, having a shower, driving the car, reading, etc, are good times to get downloads. It's not that you're being more joyful. It's more because you're *thinking less* and because you're thinking less, you're being less negative. Yes, really! And yes, this applies to *all* humans! Imagine the downloads you *could* achieve in a constant state of joy!

I don't get downloads from any kind of spiritual practice such as meditation, chanting or holding a crystal. Wearing a gold cross on a tunic whilst surrounded by rose petals doesn't seem to make a difference either. In fact, I struggle to meditate for even ten minutes a day. The November 2003 download happened whilst I was reading a book.

I have received some of my best downloads when I am sitting on the toilet because I'm usually in a positive, relaxed state. I won't say 'state of joy' because inside a toilet, that's just weird. But I will say, I am often happy and relaxed in the toilet and so I'm worrying less - less negativity, more downloads.

In my case, I can almost see my Soul yelling, 'Quick, throw in a download while he's away with the fairies and his mind is not worrying or stressing about stuff, otherwise we can't get a word in!' That's why my downloads happen the way they do - as large chunks of thought. Not because I'm some sort of spiritual master and can handle great bundles of wisdom but because my Soul has to wait for me to lessen my thinking. When my 'joyful' channel opens slightly my Soul throws a chunk of information into the gap! I know this because that too came in a download!

This book is a step by step description of what I saw in November 2003, as well as all the downloads I have received since that day.

WHY DID GOD CREATE THE COSMOS?

Let's start at the beginning. And I mean, at the very beginning. Let's first assume that some kind of super-intelligent, supremely powerful, conscious being was behind the construction of the Cosmos. The Cosmos is just too amazing for it to be created by chance evolution.

Even science agrees that the chances of a 'Big Bang' creating such a magnificently ordered and structured Cosmos are one chance in 10 to the power of 229. How big is that? That's one chance in 10, 000, 000, 000, 000, 000, 000, 000, 000, 000, 000, 000, 000, 000, 000, 000, 000, 000, 000, 000, 000, 000, 000, 000, 000, 000, 000, 000, 000, 000, 000, 000, 000, 000, 000, 000, 000, 000, 000, 000, 000, 000, 000, 000, 000, 000, 000, 000, 000, 000, 000, 000, 000, 000, 000, 000, 000, 000, 000, 000, 000, 000, 000, 000, 000, 000, 000, 000, 000, 000, 000, 000, 000, 000, 000, 000, 000, 000, 000, 000, 000, 000. *One chance!* Roughly, of course!

Let's assume that super-intelligent, supremely powerful, conscious being is GOD. The bigger question is: why did GOD create the Cosmos? The answer to this question is so fundamentally simple and yet so complex (which is typical of GOD) that it could be a whole other book.

Let's rewind history and go all the way back to the point *before* the Cosmos was created - a point known as pre-creation. At the point of pre-creation there was *absolutely nothing,* except GOD. There was no energy, no space, no time, no matter, no light ... nothing. Not even a Starbucks™ Coffee House. That meant that GOD was nothing and that nothing was absolutely everything. I call this 'nothingness'. This also meant that GOD could not experience anything. Why? You need contrast to create experience and there is no contrast within nothingness.

DUALITY

From a human perspective, take the colour blue for example. We could not understand or appreciate the colour blue unless there were other colours to compare it with. Otherwise, how would you know what blue was? Similarly, if it weren't for short people we wouldn't even know what a tall person looked like. If it weren't for pain, we would not appreciate pleasure. If it weren't for failure we could never fully understand success. The same applies to man-woman, up-down, left-right, hot-cold, day-night, rich-poor, fast-slow, solid-liquid, good-bad, physical-spirit, etc.

For every experience there is an equal and opposite *other-experience* that helps us recognise and appreciate what the first experience was, by comparing the contrast of the two experiences together. This is called duality. Otherwise, how would we know? The greater the duality, the greater the contrast, the greater the experience. This is called extreme-duality .

We are dualistic creatures. We *love* contrast. We cannot understand life or process experience without it. Just look at your own life for example. When do you mostly crave food? When you are hungry. When do you most appreciate a hot shower? When you are cold. When do you mostly seek peace? When you are stressed.

The process is even more significant in a group. Imagine a room filled with very well-behaved teenage boys of the same age, same backgrounds and same IQ. Imagine leaving them alone to mingle amongst themselves for a day. It would be interesting and fun at first but after a while it would get boring. Same, same, same. Now imagine what would happen if you introduced some teenage girls into the room. The room would suddenly *spring to life* because of all the extreme-duality. Duality actually *creates* life (literally in this case, if no adults were present!).

The good news is that we don't have to experience *all of our* duality, first hand. Duality can be internal or external. In other words, you don't have to lose a leg to appreciate a limb but you can certainly empathise with limbless-ness by observing an amputee. Metaphysically speaking, that's *why* that person without a limb appeared in your life in the first place.

In saying that, nothing beats a *personal* experience of duality. For example, you can observe other people's success from afar but nothing will teach you more about success than personally recovering from failure. You can watch a scary film on TV but nothing can make you understand fear until you have personally *felt* scared. You can read all about personal-power in books but nothing will make you appreciate expression until you personally tell someone to, 'Shut the %$#@ up!'

On Earth we have a huge variety of physical, emotional and mental duality creating a whole range of experience. That's how we find out who we are, what we like and what we don't like. That's how we *know thyself.*

At the point of pre-creation, there was only GOD. There was no duality. Consequently, GOD couldn't think, 'I am blue because I compare myself with something else *over there* that is yellow' or 'I am tall because I can compare myself with someone else over there who is short' or 'I understand joy because I have felt anger'.

That created a dilemma for GOD. Not a dilemma in the way that we humans experience a dilemma such as 'I can't find my car keys' or 'I'm in love with my best friend's wife', but more of a cosmically beautiful Divine dilemma. The dilemma was that GOD could not fully experience the Divine Magnificence of *Herself.*

I have to give credit to Neale Donald Walsch for that phrase and the concept that he so beautifully describes in his book 'Conversations With God : An Uncommon Dialogue (Book 1)'. When I first read that statement, I must have stared at the ceiling (I was lying in bed reading) for almost an hour trying to comprehend the yooogeness of it.

THE ILLUSION OF GODLIKENESS

To solve the dilemma of GOD not being able to fully experience the Divine Magnificence of *Herself,* GOD decided to create something that was *not* GOD, so she could have something to compare Herself to. In other words, if GOD was blue, She might create something that was yellow and then observe that yellow-ness from afar, in order to give Herself a fuller and more expanded view of the Divine Magnificence of Her own blue-ness.

Creating not-GOD is a real problem, when everything, everywhere and everywhen *is* GOD, because you cannot have *not* GOD. But, you can have the *illusion* of not-GOD.

From a human perspective, illusions are something that are not-real and not-real can easily exist within the imagination of the mind. For example, we can't jump off a building and fly but we can certainly imagine it, in the mind. We can't have sex with perfect strangers but we can imagine it, in the mind (yes you have!). We can't go backwards or forwards in time but we can easily relive the past and ponder the future - we can imagine it, in the mind.

That's rather convenient because GOD is a mind. That's *all* GOD is. GOD is nothingness and nothingness is pure mind energy. GOD doesn't have arms and legs or wear a gold robe. GOD is *only* mind. Imagining is all GOD *can do*. So that's what GOD did. GOD imagined the illusion of not-GOD in the same way that we would daydream in the park.

WHAT IS NOT-GOD

So what is not-GOD? Not-GOD is not a bad thing as in evil or unholy. On the contrary, not-GOD can actually still be very GODlike and the best way to understand that conundrum is to imagine that not-GOD is not-gold.

Let's imagine GOD is 100% pure gold. 100% pure gold does not exist on Earth. There is always some impurity in gold. Even the finest gold, known by the Chinese as 'exact gold' or 'Chuk Kam', still presents a 1 in 10,000 impurity and is 99.99% pure gold. In that regard, the gold that we're familiar with is not-gold. It's very *gold-like* because it's so close to being 100% pure gold, but it's still not-100% pure gold and so it's *not-gold*. It's the *illusion* of gold.

And the illusion of gold can stretch much farther than 99.99% gold. Gold is too soft to be made into fine jewellery. Metal alloys such as silver, copper or zinc are mixed with gold to make it stronger and more durable. The purity of gold is expressed in carats - 24 carat gold is 99% gold with a very tiny amount of metal alloy impurity. 18 carat gold is 75% gold with a 25% impurity of metal alloys. 9 carat gold is 37.5% gold with a 62.5% impurity of metal alloys. The illusion of gold can then be anything from 24 carat to 9 carat gold.

Carat Purity	Gold Content
Chuk Kam	99.99%
24 carat gold	99.0%
22 carat gold	91.6%
18 carat gold	75.0%
14 carat gold	58.5%
9 carat gold	37.5%

(Table) The impurity of gold.

It may all be not-gold but it's all still pretty nice because we value the gold, not the impurity. The illusion of not-GOD is very similar. GOD does not exist within the Cosmos because the Cosmos is not-GOD. In human terms, the impurity within the illusion is negative thought. The 'gold' is positive thought. The entire illusion may be not-GOD but it's all GODlike because we value the 'good in all people' (positivity) versus the impurity of their negativity. In that regard the illusion is an illusion of GODlikeness.

For example, Buddha and Jesus may have had 99% positive thoughts with a very small 1% level of negative thought (they were 'human' after all) and would represent 24 carat gold. Great leaders and world-changing individuals may have had 75% positive thought and 25% negative thought and would represent 18 carat gold. Most 'normal' people are happy most of the time, with 58.5% positive thought and the impurity of 41.5% negative thought and would represent 14 carat gold. Criminal, deceptive and super-serious people may have a very low 37.5% level of positive thought and a very high 62.5% level negative thought and would represent 9 carat gold.

In each example, each person contains some negativity as impurity, but they're all GODlike (an illusion of GOD), with the potential to be 24 carat gold as they reduce their impurities. GOD then observes all the different levels of good and impurity for an expanded viewpoint of Her own Divine Magnificence and that observation is yoooge!

THE ILLUSION IS BIG

That illusion of GODlikeness is big. It's yoooge! To better explain just how big, let's put a mathematical perspective on that. The illusion is anything between 0-99.99% GODlikeness. Let's start with 99.99%. Unlike gold, the 99.99% value of GODlike purity continues way past .99, because mathematically speaking, not 100% is actually 99.999999999 … % *recurring*. The .999999999… bit continues on forever.

For example, you can have 99.99% GODlikeness. It's still not 100% pure GOD but it's very close. You can have 99.9999999% GODlikeness. It's still not 100% pure GOD but it's very close. You can put a trillion number nines on the end of 99.9999999 … %. It's still not 100% pure GOD but it's very close. There is no end to the number of nines you can put on the end of 99.999999999 … % recurring and so 99.999999999 … % recurs for infinity. It never ends. Now that's yoooge!

99.999999999 … % recurring continues for infinity so it gets very close to GOD *infinitely* but it is still a tiny, tiny, tiny bit not-GODlike. This creates the *illusion* of GOD within an illusion of not-GOD. Ponder that for a moment!

A similar thing happens at the other end. You cannot have 0% GODlikeness, just like you can't have 0% gold. If you took all the gold out, it wouldn't be gold, it would just be a metal alloy. If you took all the GODlikeness out of the illusion it would be no-god and that's impossible, because everything is GOD, even the illusion of not-GOD. Where's that harmonica, 'pssh, psshik, pssh, haarssh, pssh, psshik, pssh … f$#@!'?

It's as if GOD Herself cannot even imagine no-god, even within an illusion of not-GOD. That means that 0% GODlikeness cannot exist and so 0% is actually 0.0000000000 … % recurring and at some point *all the way* down the infinite end of this recurring set of zeros you have a 1 to make 0.0000000000 … 1%. There is no limit to the amount of zeros you place between the 0.0 and the final 1. This means that 0.0000000000 … 1% also recurs for infinity and never ends. Now that's yoooge!

0.0000000000 … 1% continues for infinity so it gets very close to no-god *infinitely* but it is still a tiny, tiny, tiny bit GODlike creating the *illusion* of no-god within an illusion of not-GOD. Again, ponder that for a moment! That's yooge!

When you combine the two infinite ends together, you get the infinite illusion of GOD and the infinite illusion of no-GOD, with every possible combination of the two illusions in between. Now that's even yoooger! Or is it? No, it's even yoooger-er than that because that infinity is not linear. It's not a piece of string that's infinite at both ends, it's a piece of string that's infinite in every direction, in every dimension, in every universe, in every part of the Cosmos. It's … it's … I don't know what it is because I can't comprehend it!

To try to comprehend a fraction of just how big that is, imagine infinity on a linear scale. Imagine a car garage with a big, green, metal garage door. Now imagine that door is infinite. Imagine that door has *infinite* height and width (linear X and Y infinity). That means the garage door extends up-down *forever* and left-right, *forever*.

Its infinite size would mean you'd never be able to 'stand back far enough' to view it from your limited human perspective! Not only would you not be able to see it, you wouldn't even know it was there!

That's only infinity in two dimensions - height and width. Imagine infinity in *three* dimensions! Imagine the internal car garage space having infinite height, width and *depth*, so that it extends up-down *forever*, left-right *forever* and back-forth *forever* (linear X, Y and Z infinity). Where would the garage end, where would it start? Again, how would you see it?!

That covers infinity in three dimensions but just because we humans can only perceive reality in three dimensions, doesn't mean there are *only* three dimensions! The Cosmos comprises *multidimensional* infinity, co-creating itself as *multidimensional* infinity. There is an infinite number of 'garages'. Try and imagine that! You can't. No human can! Consider also that attached to these dimensions are minus or 'mirrored' dimensions (-1, -2, -3 etc) and zero dimensions (part of zero point theory) and even supposedly a minus zero (-0) dimension for all you mathematicians to ponder. Now that's yoooge!

If you gave people some gold, some alloys, a ceramic crucible and a blow torch and said, 'Blend gold and alloy together to make a metal that is gold-like', they would produce a huge variation in what people think is gold-like. The wealthy might try and create 'Chuk Kam' because that's their idea of gold-like. The average person might create the most common 18 carat gold. The poor might melt a gold alloy into 9 carat gold and say that's gold-like. The devious might pocket most of the gold and make a gold-like alloy of 99% copper. Given enough people (infinity), you would eventually be presented with *every possible variation of* gold-likeness.

Similarly, the illusion will do the same. You could have 99.997665% GODlike, 75.22996% GODlike, 69.45678% GODlike, 33.33333% GODlike, 1.55678% GODlike, 0.000004567% GODlike, etc. The difference being that the illusion of GODlikeness has recurring infinity at both ends and so there will NEVER be an end to the possible variations of GODlikeness. The illusion of not-GOD - being the illusion of *what GOD is*, the illusion of *what GOD is not*, and all possible variations of the two in-between - will expand and continue *infinitely*. This is what GOD is observing.

SUMMARY

- In the beginning there was just GOD.
- GOD was absolutely everything and that was absolutely nothing.
- GOD had no duality and so had nothing to compare Herself to and so could not fully experience Her own Divine Magnificence.
- God decided to fix this dilemma by creating something that was not-GOD.
- You cannot have not-GOD if you are GOD since GOD is everything, everywhere and everywhen.
- So GOD created the illusion of not-GOD.
- The illusion of not-GOD is everything that is not-GOD, from 0.0000000000 …1% GODlikeness to 99.999999999 …% GODlikeness.
- The illusion has recurring infinity at both ends and so there will NEVER be an end to the possible variations of this GODlikeness - The illusion of what GOD is, the illusion of what GOD is not, and the illusion of and all possible variations of the two in between.
- The illusion of not-GOD will expand and continue *infinitely*.
- This is what GOD is observing.

GOD IMAGINES 'I AM THAT I AM'

GOD imagines the illusion of not-GOD by thinking one single thought, otherwise known as the original thought of GOD. Depending on which spiritual text you read, the thought was, 'I shall be that I shall be' or 'I am that I am'.

What does it mean? It means, 'I am GOD that I am not-GOD'. If you expand on that it becomes, 'I am GOD and I choose to create something that is not-GOD (0-99% GODlikeness) so I can observe all facets of that GODlikeness, in order to understand my own Divine Magnificence'. In simpler terms, 'I am blue therefore I create yellow in order to compare the two, to experience the fullness of *what blue is*'.

I asked for a download to explain the closest English translation (Aramaic does not always translate well into English) of the original 'I am that I am' thought and got, 'I shall co-create all things in the image and likeness of me, and observe all things, free, as they become of me'.

What's interesting is one of the most famous verses in the old testament of the Bible (Exodus 3:14) states that the name given to Moses when he asked for the original name of god was 'ehyeh asher ehyeh' (Hebrew: היהא רשא היהא), which literally translates to 'I shall be that I shall be'.

The King James Bible and most other English Bibles have over the centuries translated that to 'I am that I am', although many of these Bible translations have footnotes stating that it was most probably 'I will be that I will be'. I prefer 'I am that I am' because of its power and simplicity and so prefer to use that phrase in this book.

THE THOUGHT CREATES ... NOTHING

It's well known that our thoughts are powerful enough to influence and direct the energy of the quantum field. Thoughts create, attract and manifest events into our lives - good or bad. Your life now is a direct reflection of what you think about, most. Our thoughts can even collectively influence the weather systems and geography of this planet!

So you would assume GOD's thoughts would kick arse! You would assume 'I am that I am' would have created mind-blowing stuff, maybe even the Cosmos and everything in it? Actually ... no! The thought created nothing. For centuries religions have assumed that GOD created the Cosmos and everything in it. Heck, GOD is often called 'The Creator'. Well (hands on hips, pouting with indignation), I'm here to tell you that GOD *didn't create* the Cosmos.

GOD is pure mind-energy existing as nothingness and nothingness is perfectly still, like a mountain pond at dawn. Nothingness has no movement, no action and no outcome. That means GOD's thoughts have no movement, no action and no outcome. In other words, GOD's thoughts have *no vibration* and vibration is needed for creation. GOD's thoughts are amazing but they're not creative and so the thought 'I am that I am' did nothing!

A DIRECT EXPERIENCE OF GOD

If GOD does not do anything then who created the Cosmos? We'll get to that in the next chapter. Before we do, it's best to explain what GOD does. The only thing GOD 'does' is observe. GOD is the Grand Divine Observer. GOD just watches stuff. This is the concept of *being*.

Being is feeling. If you're angry then you are a human *being* angry. If you're joyful then you are a human *being* joyful. GOD is *being* love and She does so by observing love so She can *be* love. We can do the same. If we observe a mother holding her new-born baby, we can feel love in our hearts. We didn't *do* anything *as such,* since the observation was passive. It had no movement, no action and no outcome. Yet we can feel love in our hearts just by observing the love of another and hence we can *be* love, by observing love.

That's what GOD does, just on a much bigger scale. GOD observes *all love* from *all beings* from *all time,* to *be* love! This is how GOD *is* love. Not so much because GOD *is* love (She is but let's not confuse matters) but because She observes all love to *be* love!

All love from all beings from all time is a lot of love. Just try and understand the yooogeness of *that*! Love is amazing. We've all felt it - a child's love, romantic love, friendship, and companionship. We all know that feeling in your chest that can be so overwhelming, so complete it can make you cry tears of joy. We chase it, long for it and spend billions of dollars trying to express it. Centuries of art, literature and theatre have been inspired by love and just about every Number One hit song has been a love song (or a break-up song).

Love is the most amazing and the most beautifully intense of all human experiences. And it's not just limited to Valentine's Day flowers or hugging people. The true definition of love also includes the powerful and wise derivatives of joy such as courage, abundance, success, fulfilment and imagination. In short, love is *everything* that *feels good*.

If you want to have a direct experience of GOD then all you have to do is try and imagine what it would be like to *feel* all love from all beings from all time! Take your own life for example. Try and recall all the love expressed to you by your parents, from the moment you were born until now. That's a lot of love - nurturing, care, guidance, compassion, etc. Sure, it's *never* perfect and we can all remember days when we wanted to wrap our parents up in a box and post them to a foreign country, far, far away. But we're just focussing on the love here.

Similarly, imagine a lifetime of love expressed to you by your brothers and sisters (once again, focus on the love and not the cardboard boxes). Imagine all the love expressed to you by your relatives, role models, friends and colleagues. Imagine all the romantic love from dating, romancing and making love. Imagine all the moments of self-love - every act of courage, every success, every creative expression, every moment of knowing and all the moments of joy.

Then imagine all those times *you* expressed love by helping others in some way. Imagine all your moments of generosity, forgiveness and compassion.

Then imagine all the prayers, healing and good wishes and all the love expressed to you from the *unseen* world including your Soul, Soul group, Angels, Masters and GOD.

Now combine all that love into one gigantic feeling. Now *that's* a lot of love! But that's *just you*! Now do the same exercise for every human-being on the planet. Imagine the love expressed by trillions and trillions and trillions of loving thoughts, actions and feelings, by billions of people over thousands of years. Now bypass space and time so you can feel all the love from the past *and* all the love from the future, including the flow-on effect, karma and evolution *behind* that love … for infinity.

Now multiply that by all the trillions of galaxies in our universe, each containing trillions of stars. Now multiply that by infinite 'Big Bangs' creating infinite multidimensional universes. Now take all that love from all beings from of all time and compress that feeling down into a super concentrated single point that's so dense, it produces infinite heat and light.

Imagine feeling *that much* love and at that *intensity* for infinity. That's what GOD is observing and feeling and being. That's how GOD *is* love.

DOES GOD OBSERVE THE BAD STUFF?

That's all GOD ever 'does' - observe you, so She can be love. This is Her greatest joy and Her only task. That's *why* GOD co-created (not created but co-created, which is explained later) the Cosmos. Wouldn't you if it felt that good?

That's also the foundation of gratitude. Gratitude is the feeling felt by GOD as She appreciates the opportunity, assisted by every being in the Cosmos, to be able to feel this amount of love at this level. From a human perspective, that's why gratitude feels so amazing because gratitude is such a GODlike thing to do.

But this does beg the question: If GOD only observes us, then does GOD observe all the negativity as well? The short answer is 'no'. The long answer is 'yes' since *we are all one*. But let's stick with the short answer.

GOD wants us to feel love. GOD knows the only way humans can understand, create and seek love is by experiencing the opposite of *not-love,* which is negativity. Negativity makes us feel *bad* and whenever we feel bad we immediately crave and seek happiness - the highest form of which is love. In other words, negativity makes us seek love. The greater the poles of extreme-duality, the more urge we have to seek love. That's why we live in a world of extreme-duality.

That's what's meant by GOD's unconditional love. Not so much because GOD loves you without conditions (She does) but because GOD cannot impose conditions or restrictions on the extreme-duality *required* to ensure humans understand, appreciate and so seek love - no matter how good or bad those experiences are.

Fortunately, the Cosmic process naturally filters out negativity. GOD may observe the 'bad stuff' but She really only *experiences* the good stuff, as all love from all beings from all time! GOD only observes the gold, not the impurity.

GOD DOES NOT DO ANYTHING

The concept that 'GOD does not do anything' does to religious people, especially Catholics, what pepper spray does to people with sensitive facial skin. They place their hands over their faces, call out god's name, or Jesus or Mary, fall to their knees and gasp. It upsets them greatly, especially if they've spent their life avoiding meat on Fridays, divorce, contraception, guilt-free sex or being gay.

Why? They thought they couldn't do those things because they believed that GOD would punish them if they did, for being 'naughty'. But GOD doesn't punish people for their sins because GOD doesn't do anything. All GOD does is observe all love from all beings from all time, regardless of how much meat-eating, partner-leaving, condom-flinging, guilt-free sex and gayness you indulge in. That observation is passive non-interference. It's just watching. Not meddling, controlling and judging. She is just *being* love, by observing all the other beings, *being* love.

That aside, GOD is observing the illusion to have an *impartial* experience of what GOD is. She can't interfere with that, otherwise She's interfering with the outcome of Her own observation (quantum physicists, take note!). You get the same effect if you play chess by yourself. It's boring, unchallenging and joyless because you know what the next move will be, because you're the one making the moves. You're interfering with the outcome of your own game.

GOD does not punish people for their sins. Sure, there are ramifications for bad deeds but it's self-inflicted punishment. Your negativity co-creates a world around you that attracts other negative people and events into your life. You get to *feel* the ramifications of that negativity, first-hand as self-inflicted misery, disease and karmic causality - for as long as it takes you to change. That's 'Hell on Earth'. But you co-create it, yourself. It's nothing to do with GOD and GOD has nothing to do with it.

GOD does not sit in judgement. GOD does not get angry. GOD is not a vengeful GOD. GOD does not worry if you have sex with people, with guys or girls, or both at the same time! GOD does not make stuff happen. GOD does not flood your city, cause earthquakes or send plagues. GOD does not favour those who follow Her and reject those who don't. GOD does not create stone tablets with rules on them. GOD did not re-write the Modern Bible. There is no 'GOD's Word' or 'GOD's Way' or 'In the Name of GOD'.

These concepts are not only ridiculously human but they're not even remotely close to the passive, unconditionally loving, purely observational nature of GOD. In addition, they are all forms of *doing* and GOD does not do anything. GOD does not want anything. GOD does not expect anything. GOD does not need anything.

That pretty much wipes out every single thing that humans have *blamed* god for. So for Christians – GOD does not do anything. GOD only observes all love from all beings from all time! For the non-believers, GOD never did anything; you're going to be OK for not believing in Her; She doesn't care because She only observes all love from all beings from all time!

WHAT IF GOD DID INTERFERE?

Still find it hard to believe that GOD does not do anything? If you do (jeez you're stubborn!) then just imagine if GOD *did* decide to do things. Imagine the consequences of GOD suddenly deciding to meddle in your affairs and change things – at a whim.

For example, imagine if you studied for several years to become a doctor and suddenly GOD decided to turn you into a male stripper – and you're a woman!

Imagine if you spent years building the perfect dream home and GOD suddenly decided your new house was just a materialistic-security-substitute for a lack of self-love and decided to turn your house into a tent, in the wilderness, surrounded by nasty beasts, so you could learn first-hand about real-life security issues.

Or what if GOD suddenly decided that instead of you observing the duality of suffering from afar you would be better served experiencing duality first hand, so you awake as a three-year-old Sudanese boy dying of cholera?

On a much smaller scale, what if *we* interfered with people's lives, just as many believe GOD does. Imagine if we decided to 'play god' at a local children's soccer match. What if, instead of observing our child stumble and fall as she tries to kick a goal, we run onto the pitch, knock over the other little kids and kick the goal for her because *we* decided that was best?

That would be irresponsible, highly unfair and somewhat dangerous. You would simply be interfering with a child's learning, adventure and play. Good parents quietly *observe* their children play from the sidelines, with love in their hearts. They observe them, they love them, they feel love and so they're *being* love. GOD does exactly the same thing; except I'm sure She'd probably forgo the soggy sandwich and luke-warm cup of coffee.

When you think about it and I really mean *think about it,* who do you think is the best person to decide what's best for you as an adult? It should be you, of course, because that's why you're here. You're here to make good choices *and* bad choices, because the duality of it all helps you find out what you like and what you don't like. That's how you uncover your purpose in life. That's how you find out *who you are*. That's how you change, grow and evolve. That's how you find love.

GOD can't do that *for you,* otherwise there's a really good chance you could end up being a male stripper, in a tent, with cholera! Other people can't decide for you - history has shown that's never worked. The worst advice usually comes from the ones who love you the most. Mainly because they love you and they think they know what's best for you. But they can't possibly know what's best for you, because they're *not* you, on *your* path, living *your* life lessons, within *your* world. So your parents, partner or family cannot make choices for you.

That only leaves, you. The only person who should ever decide what's best for you, as an adult in this world … is you.

SUMMARY

- GOD created the illusion of not-GOD with one single thought 'I am that I am' - 'I am GOD that I am not-GOD'.
- The thought did not do anything because GOD does not do anything.
- The only thing GOD does is observe.

- Just like we can observe things and feel love and hence *be* love, GOD observes *all things* to feel and hence *be, all love* from *all beings* from *all time,* to *be* love!
- This is how GOD *is* love.
- GOD observes the bad stuff but only experiences the good stuff (love) because GOD only observes the gold, not the impurity.
- GOD does not do anything because it would interfere with the outcome of Her own observation.
- GOD does not punish you for your sins.
- We humans punish ourselves via the self-inflicted misery, disease and karmic causality co-created by our thoughts. That's 'Hell on Earth'.
- GOD doesn't do anything. Others can't make choices for you - especially those who love you the most. The only person left is you. You are the only one that should ever decide what's best for you, as an adult in this world.

'Through our eyes, the universe is perceiving itself. Through our ears, the universe is listening to its harmonies. We are the witnesses through which the universe becomes conscious of its glory, of its magnificence.'

Alan Watts

THE ONE

Back to the question: If GOD does not do anything then who created the Cosmos? GOD is pure consciousness so to understand the creation of the Cosmos you have to understand consciousness. The principle of consciousness is probably more misunderstood than flat-pack furniture instructions.

Firstly, I like to avoid words in this book ending with 'ness' because we end up with a discussion about the 'isness' of the 'Oneness' of the 'consciousnesses' of the 'allthatisnessesss ... ess'. Consciousness is one of those words. Secondly, the word 'consciousness' gets all mixed up with 'the brain' and 'thinking' or 'being conscious and awake', and consciousness is none of these. Instead of the word consciousness, let's use the word 'mind' since mind and consciousness are the same thing.

THE MIND

GOD is a mind. GOD is the source of all things. Therefore, everything, everywhere and everywhen is mind. What is mind? There are three elements of mind: the mind itself, the potential within the mind and thought.

The mind does not think as such. It can be viewed as more of a container that holds potential. That container is infinite and forms part of the Cosmos.

Potential is an electromagnetic *force* that urges thought. Just like electricity urges an electric motor to spin and magnetism urges two magnets to attract or repel each other, electromagnetic force urges the brain to think in a certain way.

For example, if you have the potential for success in your mind then the electromagnetic force of that potential will urge your brain to think successful thoughts. Those successful thoughts will lead to successful actions, which will create feelings of success. Feeling is *being*. Feeling successful is *being* successful. You then have the potential within your mind to *be* successful.

When you think, the thought-process transforms potential into vibration, or potential in motion. It's similar to a child who has the potential to play soccer very well. That soccer-playing-potential within the mind of the child is not a real, tangible thing. You can't collect it. You can't buy it on eBay™. It does not physically play soccer. The soccer-playing-potential is a force within the mind of the child that urges the child to think athletically and skilfully. Those thoughts then get expressed as action. The child *actually playing soccer* is a physical expression of soccer-playing-potential within its mind. The thought-action then becomes *potential in motion*.

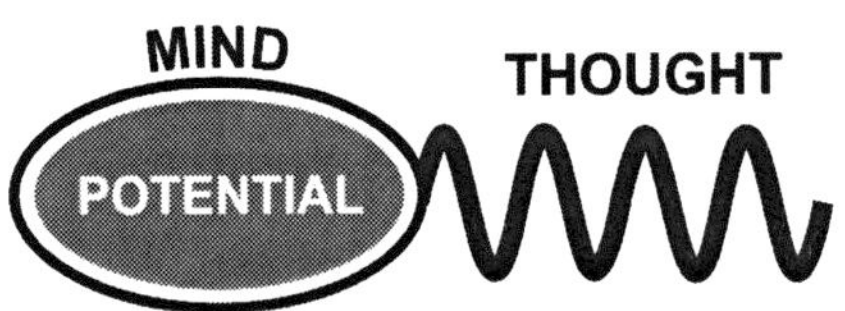

(Figure). The three elements of mind - mind, potential and thought (potential in motion). And yes it does look like a sperm (interesting!).

GOD *is* a mind and that mind is made up of two major potentials - power and wisdom. When GOD expressed the thought, 'I am that I am', the thought was more like '*I am* powerful potential *that I am* wise potential'. That thought transformed those potentials into everything, everywhere and everywhen.

Initially those potentials weren't potential in motion, like the child playing soccer. GOD is pure mind-energy existing as nothingness and nothingness has no movement, no action and no outcome. That means GOD's thoughts have no movement, no action and no outcome. GOD's original thought converted GOD's powerful and wise potential into everything, everywhere and everywhen but that everything, everywhere and everywhen had *no vibration.* It was pure, vibrationless, powerful and wise potential.

Everything originates from that single thought. That means everything, everywhere and everywhen is made up of pure, vibrationless, powerful and wise potential.

Because mind acts like a vessel for potential, if everything is made of pure, powerful and wise potential then everything must have two minds to contain those two potentials. All living beings then have two minds - a mind that contains the potential of power and a mind that contains the potential of wisdom. Next time you can't make a decision because you're 'in two minds about things' ... you *really are* in two minds about things.

This is why I refer to the mind as a 'mindset' since there are two minds, forming a *set of minds*. I only use the word 'mind' to refer to *one* of the two minds *within* the mindset. I use the term 'potential' to refer to the electromagnetic potential of the mind rather than 'the possibility of' such as, 'the potential for disaster' or 'the potential to be good at maths'.

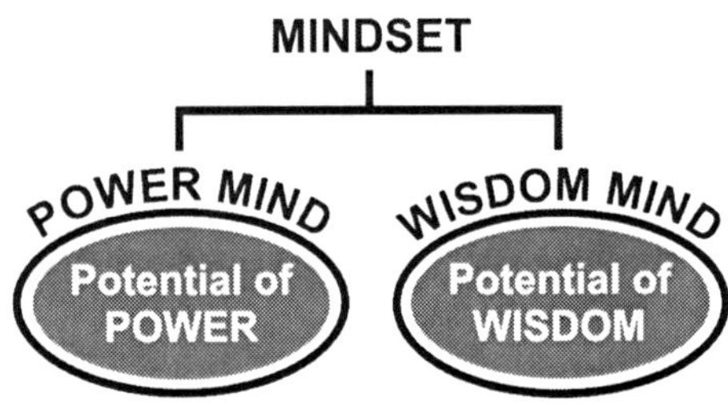

(Figure) The mindset showing the two separate minds (containers), holding the two separate potentials.

EMBRACE OR RESIST

Because *everything* is made of pure powerful and wise potential, then pure powerful and wise potential is the default potential of *all human minds*. This represents 'the good in all people', or the 'gold' in every person. That potential then breaks down further into security and capability (power), and expression and knowing (wisdom), which is explained later on.

The potential of *power* urges the brain to think powerful thoughts, which then lead to powerful actions and emotions. The potential of *wisdom* urges the brain to think wise thoughts, which lead to wise actions and emotions.

OK, if that's the case then why isn't every human thinking, acting and being (feeling), powerful and wise?

Firstly, every human is born with a limited amount of potential for each lifetime (discussed later).

Secondly, because everything, everywhere and everywhen is made of pure powerful and wise potential, there really is *only* powerful and wise potential, or *resistance* to that powerful and wise potential.

Every human has free will to either ignore or embrace their default powerful and wise potential. They can either embrace that potential, which will influence them to think positive thoughts and feel good, or they can resist that potential, which will open the doorway to negative thoughts and make them feel bad. This ensures the co-creation of extreme-duality. We can't all be perfect, otherwise we're all 24 carat gold and that's boring!

We all know what it feels like to embrace or resist potential! One feels good and the other feels bad. We've all felt that 'eerie' uncomfortable feeling when we think or do something we just *know* is not quite right. That's what resisting potential feels like. We all know what it feels like to be scared, angry, miserable or confused. That's what resisting potential feels like. We all know how wonderful it feels to be courageous, successful, fulfilled and certain. That's what embracing potential feels like.

A CLEAR DEFINITION OF GOD

Embraced potential influences the brain to think powerful and wise thoughts, which when done with a balanced mindset, creates the feeling of joy. When joy is shared with others it then morphs into the highest form of joy which is love. Love is GOD. Feeling is *being*. In other words, all humans then have the potential to *be* GODlike; a human *being* GODlike; a human *being* 24 carat gold.

If you want a clear definition of GOD, it would be this: GOD is the powerful and wise potential within the mindsets of humans that influences the human brain to think powerful and wise thoughts, which creates the feeling of joy, which leads to the feeling of love, which is a human *being* love, which is a human *being* GODlike, which is what GOD is collectively observing, from all beings from all time, so She can experience the Divine Magnificence of Herself.

Yeah right! Who's going to put that on a bumper sticker? 'Honk if you're powerful and wise potential within the mindsets of humans that influences the human brain … '.

In short, GOD is power, wisdom and love. What's really interesting is even though religions and spiritual movements across the world are so vastly different; there is this common thread amongst all religions as to the fundamental *nature* of god. Apart from mostly agreeing that god is simultaneously and infinitely everything, everywhere and everywhen, they also agree that:

- god is infinite wisdom and knows all things: god is **wisdom.**
- god is the Creator that gives life to all things: god is **power.**
- god is pure unconditional love: god is **love.**

THE ONE THINKS

That's mind. That's GOD. That's the potential effect of thought. Now that you know how that all works, I can explain how GOD co-created the Cosmos. Let's go back to the original thought of GOD for a moment.

The thought 'I am that I am' came from the mind of GOD. That means the thought *was* full of pure powerful and wise potential. Potential is an electromagnetic force within the mind that 'urges' thought. The powerful and wise potential *within* the thought 'I am that I am' made the thought *think*. So the thought, *thought* and a thinking-being is a conscious-living-being and so when the *thought* thought, it came alive. It became self-aware, as a separate living-being, with its own mindset, separate from GOD. It became the first conscious being, the first direct descendant of GOD. It became the ONE.

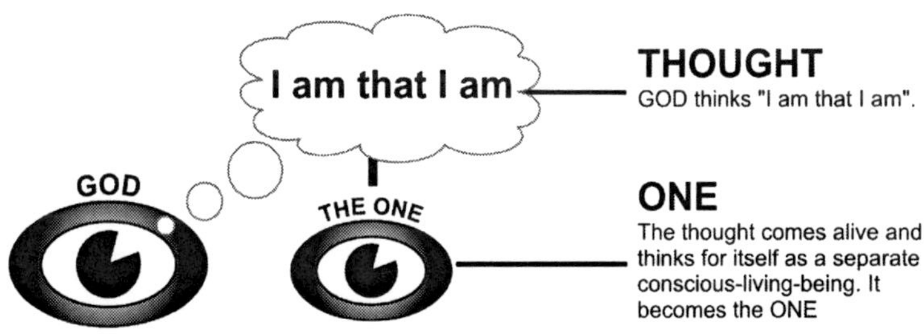

(Figure) The original thought of GOD comes alive and thinks for itself as a separate conscious-living-being. It becomes the ONE.

Being self-aware, the ONE begins to think and the first thing it thinks about is itself. The ONE contemplates its own self with questions such as, 'Who am I?' or 'What am I?' or 'Why am I here?' Haven't we all done that? So what is the ONE? The ONE is the 'living breathing' version of the original thought of GOD. It's pure, vibrationless, powerful and wise potential, direct from the mind of GOD and so when it contemplated itself it was actually contemplating the powerful and wise potential of GOD.

To do so it fully embraced all of the pure, powerful and wise potential within itself in order to think thoughts powerful and wise enough, to process the power and wisdom within itself (being the same as GOD). It was basically having a good ol' session of 'Wow, I rock!'

Or at least it tried to contemplate its own potential. Every time the ONE attempted to contemplate itself, it could only contemplate nothingness. The potentials of power and wisdom are two completely opposite ends of a duality. Mathematically speaking, they exist on either side of zero. Wisdom is above zero and so is positive (*+ve*) and power is below zero and so is negative (*-ve*). Basic mathematics will tell you that *+ve* and *-ve* cancel each other out to zero, which is nothingness.

You can experience a similar phenomenon when using noise cancellation headphones. If you wear noise cancellation headphones on an aircraft, the electronics in the headphones record the outside jet aircraft noise (*+ve* aspect) and then instantly play that noise back to you *upside down* (*-ve* aspect) in the headset earpieces. When the *+ve* noise of the aircraft combines with the *-ve* noise in your headset, the two opposing sound waves cancel each other out to nothingness. The end result is complete silence in your headphones (nothingness).

Every time the ONE tried to contemplate its powerful (*-ve*) and wise (*+ve*) potentials, those potentials cancelled each other out to nothingness. That meant the forces behind its thoughts were nothing and so it could only contemplate nothingness. That's not surprising really, since it is the original thought of GOD and GOD *is* nothingness (this was also GOD's original dilemma).

Nothingness is a great idea during a long haul flight in economy class surrounded by crying babies, loud night-talkers or the one that coughs incessantly. But imagine *contemplating* nothingness. Well, you can't because humans cannot contemplate nothingness. So instead, from a human perspective, imagine staring at a blank white wall, for a million years, no break, in total silence, as a deaf blind person. That's a very insignificant human equivalent of contemplating nothingness. Imagine the Cosmic version? Boring!

THE ONE BECOMES TWO

After a few uncountabillion eons of complete boredom the ONE decided to split into two minds, separating the potentials. It split into a power-mind which contained the potential of power, and a wisdom-mind that contained the potential of wisdom. It could then become its power-mind for a moment and let its powerful potential influence it to think powerful thoughts. It would then shift over to its wisdom-mind for a moment and let its wise potential influence it to think wise thoughts.

It could do this, again and again, rapidly shifting back-and-forth between its two minds, thinking with power, then wisdom, then power, then wisdom, all the while avoiding the +ve and -ve potentials *behind* the contemplation nullifying its thoughts to nothing.

When the ONE split its potentials into two separate minds, it also split its pure, powerful and wise potential *energy* into two separate energies. Originally, its pure, vibrationless, powerful and wise potential was joined as electromagnetic energy. When it split into two minds, it split that powerful and wise electromagnetic energy into powerful electrical-potential and wise magnetic-potential.

Each mind then ended up with its own separate force to influence thought. This is what gave each mind its characteristics. The *electrical*-potential within the ONE's power-mind urged the ONE to think thoughts of ***security*** (survival, courage and abundance) and ***capability*** (skill, success and willpower). **This potential/thought combination gave the power-mind a male characteristic to become the male mind.**

The *magnetic*-potential within the ONE's wisdom-mind urged the ONE to think thoughts of ***expression*** (personality, individuality, fulfilment) and ***knowing*** (creativity, imagination and life-experience). **This potential/thought combination gave the wisdom-mind a female characteristic to become the female mind.**

The ONE split into male and female - the first manifestation of extreme-duality. The ONE became the TWO.

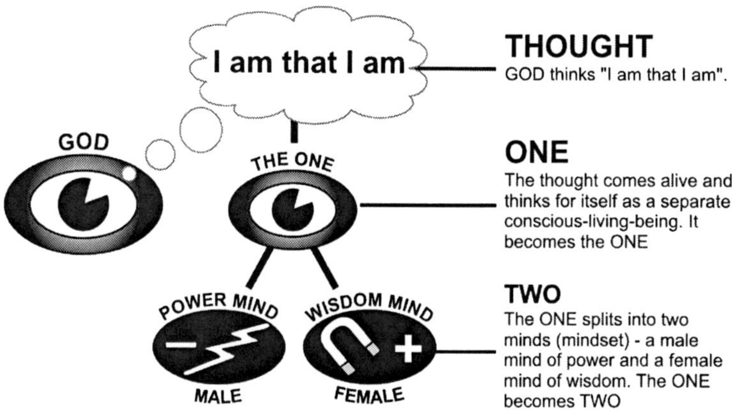

(Figure) The One splits into two minds (mindset) - the male mind of power (containing electrical-potential) and the female mind of wisdom (containing magnetic-potential). The ONE becomes TWO.

THE PULSE WAS FAST!

The TWO pulsed back-and-forth between its two minds, to allow each force within each mind, to urge each set of thoughts, so it could contemplate its powerful and wise minds *separately*. This created a pulse - a very, very, very fast pulse. The TWO pulsed back-and-forth between its two minds 400-400,000 nonillion times a second. Nonillion is *10 to the power of 30*, or a 1 with 30 zeros behind it.

How fast is that! A pulse back-and-forth of 400-400,000 nonillion hertz is 400-400,000 million, million, million, million, million times a second. Now that's fast. I believe that's the frequency of consciousness with the 400-400,000 range accounting for the frequency difference between power and wisdom.

How did I come up with the number of a nonillion? I settled on that number for two reasons. Reason number one is because gamma rays are the highest detectable form of electromagnetic vibration and National Aeronautics Space Administration has reliable detections of very high energy gamma-ray radiation from the Crab Nebula that have extended up to about 1,000 yottahertz (10 to the power of 27 hertz). Diffuse gamma-ray emissions are expected to extend up to at least 1,000,000 yottahertz (10 to the power of 30 hertz). Gamma radiation is produced from high energy states such as lightning strikes, atomic nuclei (gamma decay) and exploding stars. Those electromagnetic elements are fairly 'Cosmic' so I *assumed* the frequency of mind must be just above that frequency - 400-400,000 nonillion hertz is [4x10 to the power of 32] - [4 x10 to the power of 35]. It's in the range, at least.

Reason two, the main reason, is that we are holographic replicas of the Cosmos. Whatever happens in us also happens in the Cosmos. If you want to unravel the Cosmos, which is made of mind, you only have to unravel the human mind.

Humans have been universally fascinated by one major, common element for thousands of years - music. Music (drums, chanting, singing and instrumentation) dominated culture before history was recorded. Music creates a pulse that produces a vibrational waveform we experience as sound. The mind creates a pulse that produces a vibrational waveform that we experience as reality (we'll get to that). They are connected and music vibration is simply a lower octave of thought vibration.

Music generally ranges from 20-20,000 hertz. Life, the outcome of the mindset, *currently* evolves from its lower mindset to its upper mindset in five rounds of three cycles of seven years (5 x 3 x 7=105 years). The general population doesn't seem to live past the age of 105 years because our mindsets have not evolved past this level. Because we are holographic images of the Cosmos, I calculated that consciousness must therefore vibrate somewhere in the vicinity of 105 octaves above 20-20,000 hertz. That creates a frequency range of 400-400,000 nonillion hertz. This may seem a little simplistic but life *is* as ordered as it is chaotic.

THE CREATION OF SPACE AND TIME

As the TWO pulsed back-and-forth between its two minds, it repeatedly transferred the electromagnetic forces *within* its thoughts, back-and-forth into nothingness.

From a human perspective, electrical-potential influences the brain to think thoughts of security (survival, courage and abundance) and capability (skill, success and willpower). That's a very *action now* type of thinking. It's a very 'go get it', 'have it', 'want it', 'do it', 'succeed or fail' type of thinking. In other words, the potential of power urges a very 'on' or 'off' type of thinking, in the same way that electricity either turns a light bulb on or off. It's either going to happen or it's not and this is usually decided by your level of will-**power** or personal-**power.** In short, it's power.

When the TWO thought, the electrical-potential within its thoughts of power, distorted nothingness in an 'on-off' kind of way. This created the Cosmic element of space since (cosmically speaking) you need space to go up-and-down like a switch, to cater for the 'on-off' potential of power.

From a human perspective, magnetic-potential influences the brain to think thoughts of expression (personality, individuality, fulfilment) and knowing (creativity, imagination and life-experience). These types of thoughts get stored in the subconscious mind to create memories. Your memory has the ability to either retrieve or push away memory-files (past-thoughts) in the same way that two magnets either attract or repel each other.

When the TWO thought, the magnetic-potential within its thoughts of wisdom distorted nothingness in a 'push-pull' kind of way. This created the Cosmic element of time since (cosmically speaking) you need a past, present and future to move from side-to-side (back-and-forth in time) to cater for the 'push-pull' potential of wisdom.

SPACE-TIME CREATES WAVES

The distortion of space moves up-and-down over *nothingness*. The distortion of time moves side-to-side over *nothingness*. In this case nothingness is zero and so both distortions share the same *zero*-point centre and that common zero-point re-combines the two distortions together so they *both share* space and time. This is known as space-time and *this* is where the fun begins.

Space is simple and predictable. It always goes up-and-down. Time is relative, which makes it weird! Time has to go side-to-side so that we can go back or forward in time, in our imagination. But in reality, the past has gone and we always *move forward* into the future, to allow for the expansion of the Cosmos. In other words, time can go back and forth in the mind but always moves forward in reality to create the passing of time.

When the common zero-point re-combined the distortion of space with the distortion of time, they *shared* each other's elements of space-time. Instead of the distortion of space *just* going up and down, it suddenly went up-and-down *as* it moved forward through time. Instead of the distortion of time *just* going side-to-side, it suddenly went side-to-side *as* it moved forward through time.

Take the distortion of space for example. If you point your index finger at a wall and move it up-and-down (about two times a second) it will simulate the distortion of space as it moves up-and-down over *nothingness*. Now slowly move your hand to the right *while* your index finger is moving up-and-down. This will simulate the up-and-down distortion of space as it moves through time. What shape do you notice your fingertip is tracing on the wall? Yes ... it's a waveform (see figure below).

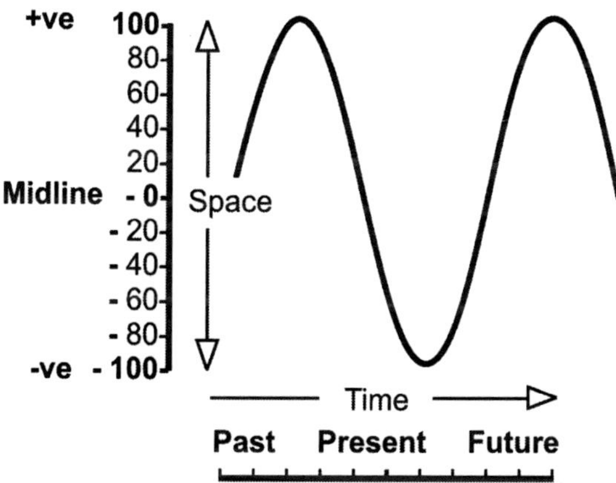

(Figure) Pulsing back-and-forth while moving forward over time to create a waveform.

RIPPLES IN THE POND

You have just created the fundamental force behind all reality, with your finger - waveforms! Waveforms create ripples in nothingness like a stone creates ripples in a still mountain pond. A mountain pond's surface is still, glassy, motionless. The splashing impact of the stone disrupts the stillness of the pond surface and causes the surface of the pond to rapidly move up-and-down. That rapid up-and-down movement creates concentric ripples in the pond that emanate out from the centre. They're waves travelling outwards in form (circles) - waveforms.

(Figure) A stone creates ripples in a still mountain pond.

Thought does a similar thing to nothingness. In this example, the still surface of the mountain pond represents nothingness and the stone represents the electrical or magnetic potential within thought. When a conscious-living-being thinks, it pulses back and forth between its two minds. The electrical and magnetic potential within its thoughts rapidly distorts the surface of nothingness.

The Cosmic-pond is weird because there are two ponds - one horizontal pond and one vertical pond. The electrical-potential behind powerful thought rapidly distorts nothingness up and down to create space. The magnetic-potential behind wise thought rapidly distorts nothingness side-to-side to create time. The common zero-point of nothingness combines space and time together so each distortion shares space-time. This converts the distortions of space and time into ripples (waves) of space and time.

In other words, the up-and-down distortion of powerful-thought co-creates ripples of space in the horizontal pond, and the side-to-side distortion of time co-creates ripples of time in the vertical pond.

Nothingness becomes a pond 'surface', rather than just nothingness, due to the sudden addition of space-time. Ripples in nothingness are made of space-time so they don't travel outwards from the centre like the concentric ripples of a mountain pond. Ripples of space-time don't have to travel through space and time. They *are* space-time and so they bypass space and time altogether and instead, are already everywhere all at once. It's as if you took a snapshot of ripples in a still mountain pond, 20-30 seconds *after* you threw the stone into the centre of the pond. The photo would show all the ripples, right to the edge of the pond. The ripples are static in that regard, yet individually they still 'do their thing!' - go up-and-down or side-to-side.

This means ripples in the Cosmic-pond are everywhere all at once, as three dimensional, concentric, spherical ripples, otherwise known as thought-waves.

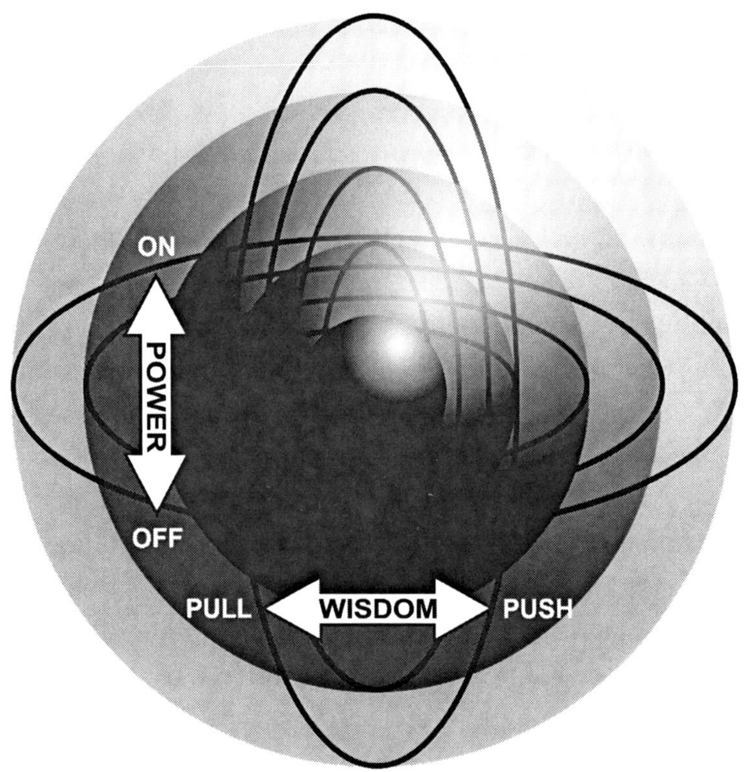

(Figure) The up-and-down distortion of powerful-thought creates ripples of space in the horizontal pond and the side-to-side distortion of time creates ripples of time in the vertical pond. Ripples of space-time are everywhere all at once, as three dimensional, concentric, spherical ripples. These are thought-waves.

Each nonillionth of a second pulse is the equivalent of another stone being thrown into the pond and so the ripples are far more intense than I could show in the figure above. The ripples would look more like the figure below, except a nonillion times more complex.

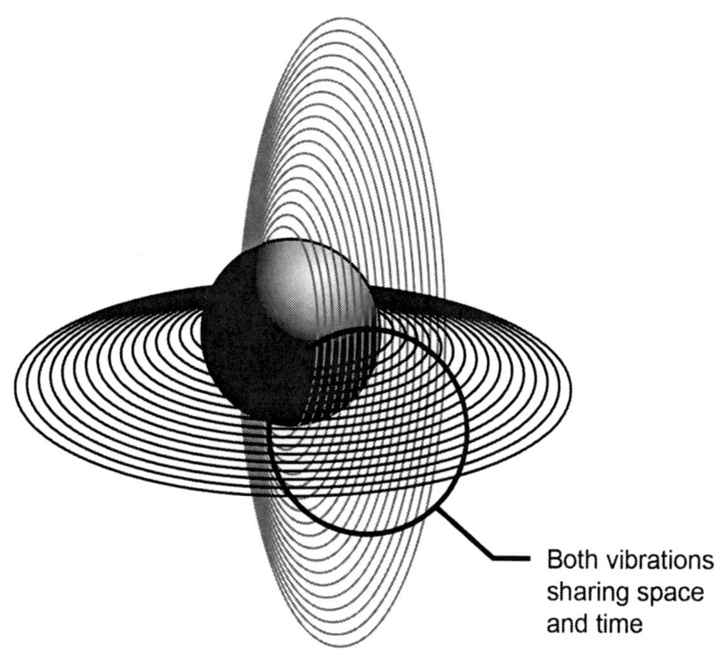

(Figure). Showing a representation of the nonillions of ripples created by thought, in both ponds, interacting to form the grid of the Cosmic-pond.

THE TWO BECOMES THREE

When the ripple of time re-combines with the ripple of space at the common zero-point, the energies of the two ripples also re-combine. The electrical-potential behind the powerful thought that created space re-combines with the magnetic-potential of wisdom that created time, to form a combined, *electromagnetic,* double ripple.

The two ripples recombine ninety degrees out of phase from each other to avoid wave collisions and cancellations. Amazingly, that double-ripple ends up being magnetic-potential through space *over* electrical-potential through time. Now I'm pretty sure that's what scientists call a 'linearly self-propagating transverse oscillating wave of electric and magnetic fields'. Try saying that with a mouth full of dry biscuits! Let's just call it electromagnetic vibration for short.

If you're a scientist, electromagnetic vibration is light (light is a pulse wave of electromagnetic energy). If you're a metaphysician, electromagnetic vibration is what makes up all reality. When nothingness is still, it's nothing. When nothing vibrates with space-time it becomes reality - something, somewhere, somewhen. If you're me, then electromagnetic vibration is the *Light* I saw in November 2003!

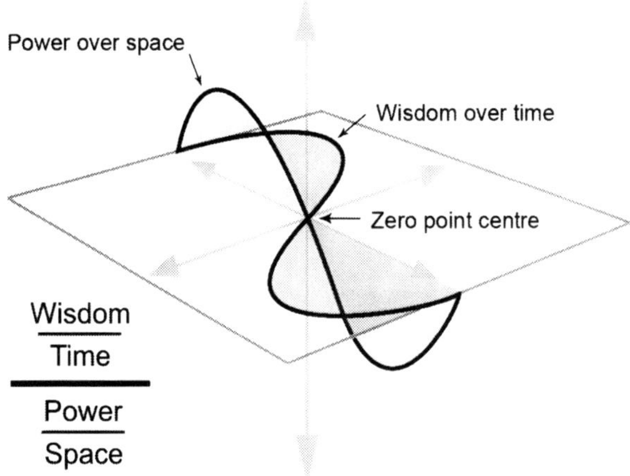

(Figure) The co-planar electromagnetic waveform of Light. To see this wave in motion go to www.yoooge.com and click on the 'Links' button, then click on the 'Electromagnetic Waveform' link.

Electromagnetic vibration then becomes the third co-created element to form the trio of power, wisdom and *vibration*. The TWO then becomes the THREE.

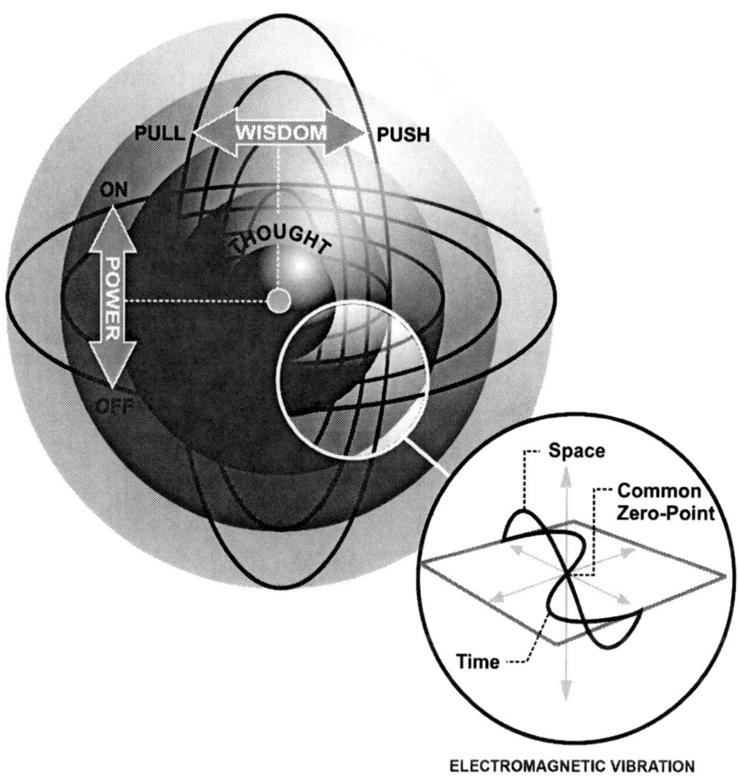

(Figure) The two waveforms of power and wisdom re-combine at the common zero-point to form electromagnetic vibration.

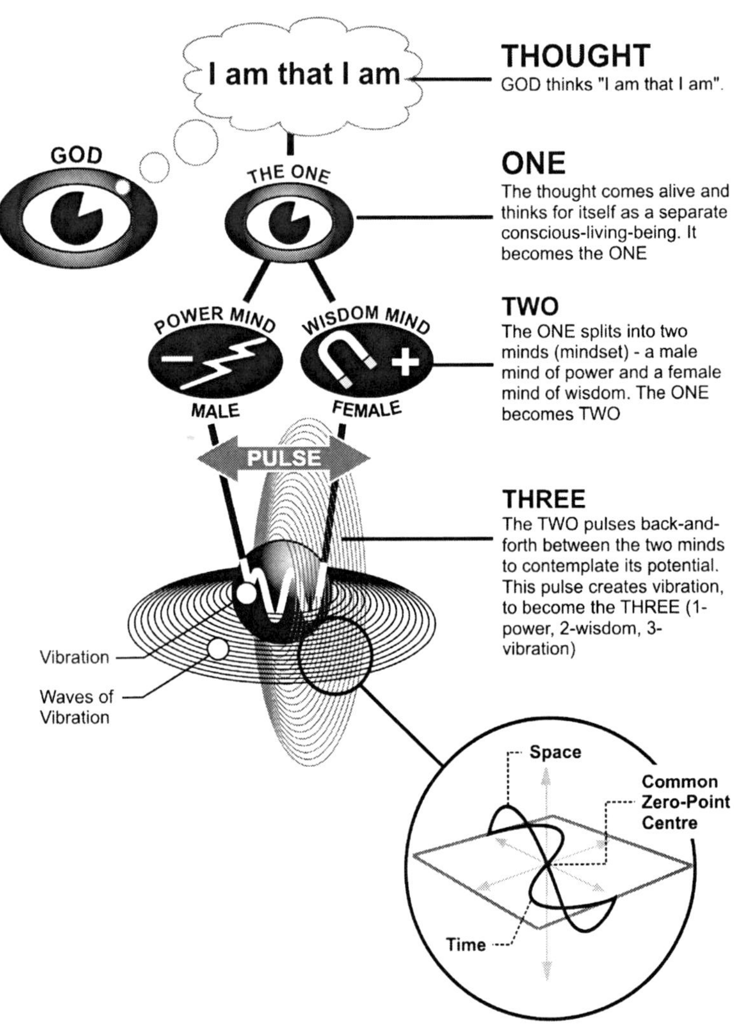

(Figure) The co-creation family tree showing the connection between GOD and the ONE, TWO and the final outcome of electromagnetic vibration - the THREE.

All reality is made up of vibration - no exception. Some people call it energy; others call it consciousness, whilst the religious folk call it god. Either way, whether it's your big toe, planet Earth, your fears, events in life, ideas in your imagination, your life's purpose, life itself or even death - it's *all* vibrational-reality. That's kind of spooky when you consider that *everything* we think of as *being real* is made of this buzzing, pulsing vibration. In other words, a house brick is more of a song than a solid object. Your career is more like a pulsing speaker-cone in a stereo than a lifetime of hard work. Your imagination is more like a buzzing fluorescent light than thoughts. Life is a buzz!

The wonderful thing about vibration is that it's made of waveforms and waveforms are spread out over space and time. That means their *+ve* and *-ve* potential values are spread out over space-time, so at no point is any *+ve* value directly opposite any *-ve* value. This allows two completely opposite aspects of a ripple to co-exist together without any of their *+ve* and *-ve* values cancelling each other out to nothingness - yet still *maintain* an energetic basis of nothingness.

This allows the mindset of any conscious-living-being to contemplate powerful and wise thought, without the potentials within those thoughts cancelling each other out to nothingness, within an illusion built from nothingness. Ponder that!

ONE, TWO, THREE ...

Confused? Hang in there! We're almost done. When the TWO pulsed between its two minds, nonillion times a second, to contemplate its powerful and wise potential, it fully embraced its GODlike potentials, producing ripples of space-time (electromagnetic vibration) the equivalent of throwing a stone the size of a star system into a pond the size of GOD, *nonillions of times a second.*

Space added dimension to nothingness, spreading nothingness up-and-down *lengthways,* like paint on a wall. Time added a past-present-future to nothingness, spreading nothingness back-and-forth *side-ways,* like butter on a sandwich. The combination of the two produced an incredible expansion that transformed everything that was nothing (nothingness), into something that was simultaneously, instantly and infinitely everything, everywhere and everywhen.

BANG!

SUMMARY

- Consciousness is mind.
- Mind is made up of the mind itself, potential within the mind and thought.
- The mind is a container of potential.
- Potential urges the mind to think.
- Thoughts are potential in motion.
- GOD is a mind and so everything, everywhere and everywhen is mind.
- GOD does not do anything so GOD's thoughts do not do anything and so when GOD expressed 'I am that I am' it turned everything, everywhere and everywhen into pure (vibrationless) powerful and wise potential.
- Every conscious-living-being then has a default level of powerful and wise potential and two minds to contain these two potentials - the mindset.
- Humans have free will to either embrace or resist that potential. One feels good and one feels bad.
- Fully embraced potential will create thoughts of power and wisdom, which creates joy, which leads to love which is a human *being* GODlike.
- The thought of GOD comes alive and thinks for itself. It becomes the ONE.
- The ONE then splits into two minds to contemplate its powerful and wise potential. This also splits its original electromagnetic force into electrical-potential and magnetic-potential. This gives each mind its characteristics producing a male mind of power and a female mind of wisdom. The ONE becomes TWO.
- The TWO then quickly pulses back-and-forth between each mind to separately contemplate its own potential.

- The electrical-potential within the TWO's thoughts distorted nothingness in an 'on-off' fashion to create the Cosmic element of space.
- The magnetic-potential within the TWO's thoughts distorted nothingness in a 'push-pull' fashion to create the Cosmic element of time.
- These two ripples then each share each other's elements of space and time to create waveforms.
- These two waveforms then re-combine at the common zero-point to create electromagnetic vibration. The TWO becomes THREE (power, wisdom and vibration).
- When nothingness is still, it's nothing. When nothing vibrates with space-time it becomes reality - something, somewhere, somewhen.
- The TWO's contemplation transformed everything that was nothing (nothingness), into something that was simultaneously, instantly and infinitely everything, everywhere and everywhen.
- **BANG!**

THE COSMOS

You can't suddenly shift from a state of nothingness into everything, everywhere and everywhen, without creating a %$#@en Big Bang! Actually it was more of a 'brrr!' than a 'bang!'

Even though the ONE is the first part of the conscious-living-being, its element of TWO did the contemplation and the THREE is the vibrational result, I'm going to refer to the ONE-TWO-THREE as 'the ONE'. Otherwise it gets too messy talking about the ONE-TWO-THREE. In addition, the ONE is original conscious-living-being that came alive and thought for itself and so it is ONE-TWO-THREE.

The thought-waves from the ONE's contemplation suddenly added 400,000 nonillion star-system-sized ripples of space-time to what was previously nothing. That's the *explosive equivalent* of operating a jack hammer in a tea cup filled with custard, except the jack hammer is hammering at 400,000 nonillion times a second and the jack hammer blade is as big as a horse. 'But a horse doesn't fit into tea cup!' I knooow. That's why it went Brrr! One moment it was still, quiet, and serene; the next there is a sudden and violent vibrational expansion. 'Brrr!'

Cosmologists have their more sensible viewpoint. They believe that something, 13.7 billion years ago, went from a single point of infinite density, infinite heat and infinite light and expanded at an extraordinary rate, many times faster than the speed of light. This expansion produced a massive cosmic explosion of unimaginable intensity and colossal proportions, creating the infinite Cosmos as we know it.

About one percent of the flickering static specks you see on your television screen are created by the energy *left over* by The Big Bang (primordial cosmic radiation). You can actually see the leftovers of the ONE's contemplation of itself on your TV screen! Who knew?! Choose a channel on your TV that's pure static, meditate on it one day and discover what happens.

Cosmologists call the point before creation, 'The singularity'. They state the singularity didn't appear in space – instead space began *inside* the singularity. Prior to the singularity, nothing existed. They state that this indescribable expansion was not like any normal explosion where matter is hurled outwards from a central point. It was more a 'suddenly everything is there' type of explosion, or an explosion of space within itself. Energy and matter being hurled in all directions is simply the turbulent result of this incredible expansion.

Cosmologists can actually pinpoint the moment when space and time were introduced to the Cosmos - a 10th of a millionth of a trillionth of a trillionth of a trillionth of a second (10 to the power of -43) after the Big Bang.

The Cosmos is still expanding today at a rate of about 75 km per second per 3.26 million years. In other words, galaxies appear to be receding from us at a rate of 162,000 miles per hour for every 3.26 million light-years farther out we look. The farther we look the faster the recession. It's known as The Hubble Constant.

THE COSMIC-POND

Contrary to the popular belief, the Big Brrr didn't create star systems and galaxies and conscious-living-beings and Starbucks™ coffee houses. The Big Brrr simply created ripples of space-time. It may well have been 'everything, everywhere and everywhen' but there was nothing in it. It was dark and empty. The ripples of space-time had nothing to interact with and interaction is needed to turn vibration into reality.

The ripples of space-time do perform a valuable service. They hold vibration together as reality. This is a good thing since vibration *on its own* is not a solid, tangible substance. It does not make a good illusion. You can't sit on pure-vibration, drive it to work or eat it any more than you can sit-drive-eat a Jimmy Barnes song.

Just like how a magnet attracts and repels iron (push-pull) and just like how electricity switches things on and off (on-off), the electromagnetic forces *within* the ripples of space-time, push-pulls and on-off's space-time into a sort of grid. That grid then organises space-time into a form, function and phenomenon. I call this grid the Cosmic-pond but it's also known as the matrix, the quantum field or the universal field.

The Cosmic-pond is what makes solid things appear to be solid. It makes the green trees appear to be green. It stops us falling through the chair and the chair from falling through the floor. It makes the air we breathe cool. It makes food taste fantastic and stops our shiny new car from atomically dismembering and flowing down the street as a pile of disconnected particles.

The Cosmic-pond also accounts for all Cosmic forces that keep scientists awake at night such as strong nuclear forces (binds protons and neutrons together to form the nucleus of an atom), weak nuclear forces (radioactive decay of subatomic particles), gravity and electromagnetic energy.

The Cosmic-pond is also *alive*! Let's not forget the Cosmic-pond is the living 'breathing' manifestation of 'I am that I am'. It's the ONE. It's alive, self-aware, has mind-blowing levels of potential and thinks for itself (all thoughts do!). The Cosmic-pond *decides* what form, function and phenomenon is required, depending on the interacting thoughts of other conscious-living-beings.

The Cosmic-pond can also account for all the hidden freakiness of the astral-world and spirit-world. It's the framework of dreams, thought, emotion and karma. It's the Light of GOD and the power of Gods. It's what gives life to a seed when you plant it in the ground and it's what leaves your body when you die. It's like 'the force' described by Obi-Wan Kenobi in the film 'Star Wars'.

'The Force is what gives a Jedi his power. It's an energy field created by all living things. It surrounds us and penetrates us. It binds the galaxy together'

Obi-Wan Kenobi, from the film 'Star Wars'.

Oh come on! I had to get one 'Star Wars' quote in. And I'm talking about the original first movie, 'Star Wars. Episode IV: A New Hope'. Star Wars was one of the first films to introduce some of the very powerful principles of Buddhist cosmology and Zen Buddhism into mainstream Western society - in the 1970's! Even the creator and director George Lucas admitted that Star Wars was heavily influenced by Eastern thought. That movie changed human consciousness on a grand scale and yet most of the Western world had no idea!

THE FUNCTION OF THE COSMIC-POND

The Cosmic-pond not only supports reality in form, function and phenomena but it also supports the co-creation of that reality.

The grid of the Cosmic-pond is flexible and organic and it behaves very much like the surface of a mountain pond. In a way, water is a solid because it cannot be compressed. But water is also fluid and flexible making it a *liquid*-solid that can morph to suit the shape of the container it's within. The Cosmic-pond has similar properties. It has a space-time framework that can make things seem solid and real but it's also fluid and flexible. It will morph to suit whatever container (form, function and phenomena) is required. We don't see that fluidity because the Cosmic-pond operates on a vibrational frequency way outside the perceptions of humans.

The Cosmic-pond supplies a medium or surface for thoughts to vibrate within (thoughts don't travel; they're everywhere all at once). When a conscious-living-being thinks, the potential within its thoughts distorts the Cosmic-pond in the same way a child's bodyweight distorts a trampoline surface downwards, like an upside-down cone. Except in this case, thought not only distorts the Cosmic-pond *down* in a cone but *also up*, as if the trampoline surface were glued to the feet of the child as he/she bounced upward.

The Cosmic-pond is then disrupted downwards and upwards, downwards and upwards, over and over again to create a 'rubbery' kind of pulse. That rubbery pulse then creates ripples in the Cosmic-pond that are everywhere all at once. Remembering, there are two ponds and so the child also bounces side-to-side. Freaky! The Cosmic-pond is also disrupted side-to-side in a 'rubbery' kind of pulse

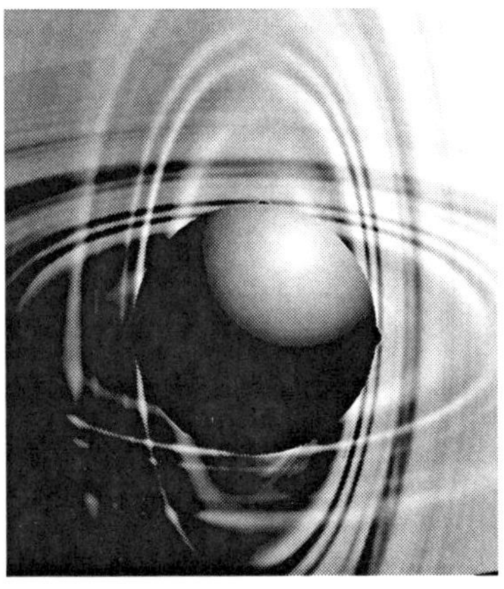

(Figure) Horizontal and vertical ripples of the Cosmic-pond.

FREQUENCY AND AMPLITUDE

Vibration is characterised in two ways, amplitude and frequency. If you compare that to a piano keyboard, frequency represents the different notes (the different keys) on the keyboard and amplitude represents the volume of each note (how hard you bang the keys).

The frequency of vibration determines the *type* of vibrational-reality. For example, ice, water and steam are the same element: water. They only differ in states because of their different frequencies of vibration - they buzz differently.

Ice is dense, heavy and solid because it has a lower frequency of vibration than water. If you heat ice in a pan the heat energy of the stove is transferred to the ice and subsequently raises the vibrational frequency of the ice and it soon turns back to water. Water is still a solid (it cannot be compressed) but a fluid solid and only because water has a higher vibrational frequency than ice. If you keep heating the pan, the water will raise its vibrational frequency even more and turn to steam. Steam is a high vibrational frequency state of water and is transparent (almost invisible), weightless and gaseous.

In this example, ice would represent the lower frequency vibrations of the *seen* physical-world - what we *can see* and touch and interact with such as objects, biology, sunlight and gravity. Water would represent the 'in-between' medium frequencies of the astral-world - what we can perceive and interact with but not quite hold in our hands such as dreams, the human aura, or your memory. Steam would represent the high frequency vibrations of the *unseen* spirit-world - what we *can't see* or interact with (not directly such as memories or dreams) such as ultraviolet light, your intuition, ghosts or your Soul.

The amplitude of vibration determines *how real* the vibrational-reality is. Vibration won't become vibrational-reality (stuff) until you turn up the volume. The more amplitude (volume) a vibration has, the bigger the up-and-down mountain peaks of the vibration are and the more real the vibration becomes. It's almost as though the vibration has to get bigger for us to 'see it'.

For example, the thought of a new car and an *actual* new car are very similar frequencies. The difference is that the vibration of the new car is much louder (has a higher amplitude) than the thought of a new car. This makes the vibration of the new car bigger, making the new car more real so that it eventually becomes physical, real-world, vibrational-reality. We can see it, touch it and drive it. The thought of a new car has a much lower amplitude and a much smaller vibration. It's not as real and so is hidden in the spirit-world as unseen vibrational-reality. We can't see it, drive it or touch it.

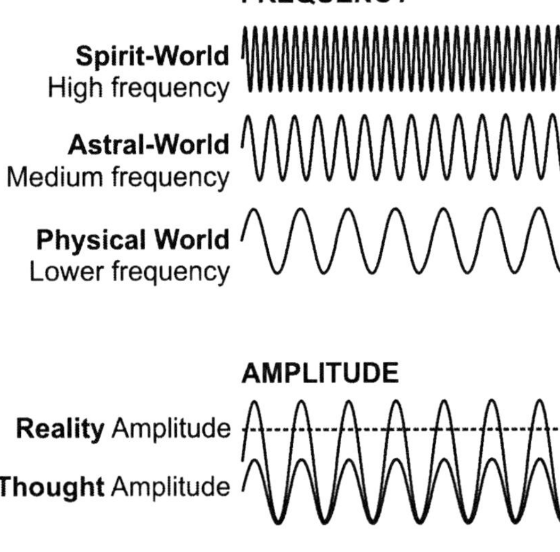

(Figure) The differences between the frequency and amplitude of vibrational-reality.

CO-CREATED VIBRATIONAL-REALITY

When a conscious-living-being thinks it creates nonillions of ripples of thought (thought-waves) in the Cosmic-pond but they don't have enough amplitude to create vibrational-reality. When those ripples interact with the existing space-time ripples of the Cosmic-pond, they clash producing points of double-amplitude loud enough to become vibrational-reality - some*thing.* That some*thing* is also interacting with ripples of space and time and so it becomes something, some*where,* some*when.* In other words, it becomes co-created vibrational-reality.

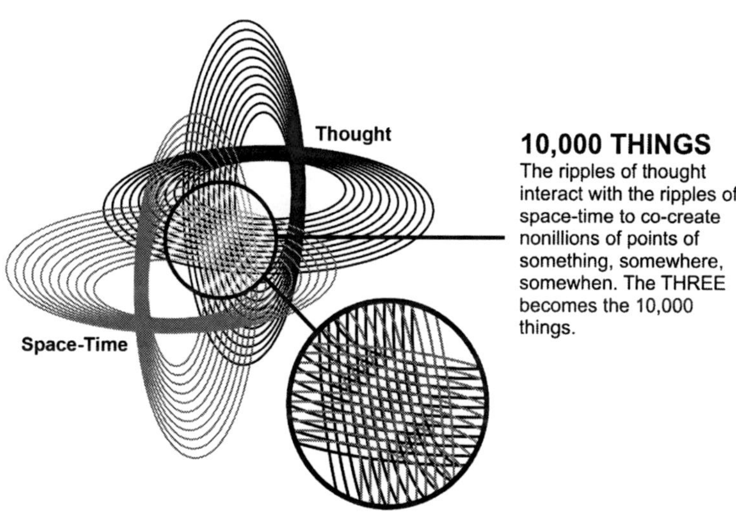

(Figure) The ripples of thought-waves interacting with the space-time ripples of the Cosmic-pond to co-create something, somewhere, somewhen.

Ripples of thought, including space-time, are everywhere all at once and so co-created vibrational-reality is also everywhere all at once, except the amplitude of that vibration reduces the further you move away from the centre. In human terms, it gets quieter the further you move away from the 'dance party', although in this case the 'Doof! Doof!' beat of the dance party is nonillions of times a second. Co-created vibrational-reality is then always louder and hence more real the closer you get to the source of the original thought - the thinker.

THE FIRST BEING

When the Cosmos was originally created, there were no other conscious-living-beings around to think. There was just the dark, empty, vibrating framework of space-time. That meant there were no other sources of vibration to interact with the vibrations of space-time (Cosmos) and create anything real. Consequently, the Cosmos decided to create some other conscious-living-beings 'to play with'.

The Cosmos created those beings in the same way that GOD co-created the Cosmos - ONE, TWO, THREE. Every thought from every conscious-living-being follows this same sequence of ONE-TWO-THREE, just on a much smaller, quieter, less dramatic scale. So I'm not referring directly to the ONE-TWO-THREE elements that made up the Cosmos, I'm referring to the ONE-TWO-THREE as a co-creative-thought-process. In addition, the Cosmos/First Being interaction that followed created what's known as the 10,000 things. The co-creative process then becomes ONE-TWO-THREE-10,000-things.

To create a being, the Cosmos *thought of the first being*, just like GOD thought 'I am that I am'. The *thought of the first being* was a very powerful and wise thought and so was *filled* with powerful and wise potential. Potential is an electromagnetic force that 'urges' thought, so the *thought of the first being,* thought. A thinking being is a conscious-living-being and so the moment the *thought of the first being* thought, it came alive. It then became self-aware as a separate conscious-living-being. It too became the ONE.

The ONE began to think for itself and the first thing it thought about *was* itself. *It* was powerful and wise potential and so it contemplated its own powerful and wise potential. To do so it had to split into two minds of power and wisdom, so the potentials behind the contemplation didn't cancel each other out to nothingness. The ONE became the TWO.

It then pulsed between its two minds, nonillions of times a second, to contemplate its power and wisdom. This pulse created ripples of electromagnetic vibration in the Cosmic-pond. The TWO became THREE (vibration).

The ripples of the THREE then interacted with the space-time ripples of the Cosmic-pond (the Cosmos) to create nonillions of points double-amplitude (something), mixed with space-time (where and when) to co-create something, somewhere, somewhen - the form, function and phenomenon of the *First Being*. The THREE became the 10,000 things. The First Being co-created itself, as the thought of itself, just like the space-time framework of the Cosmos created itself, as the thought of itself.

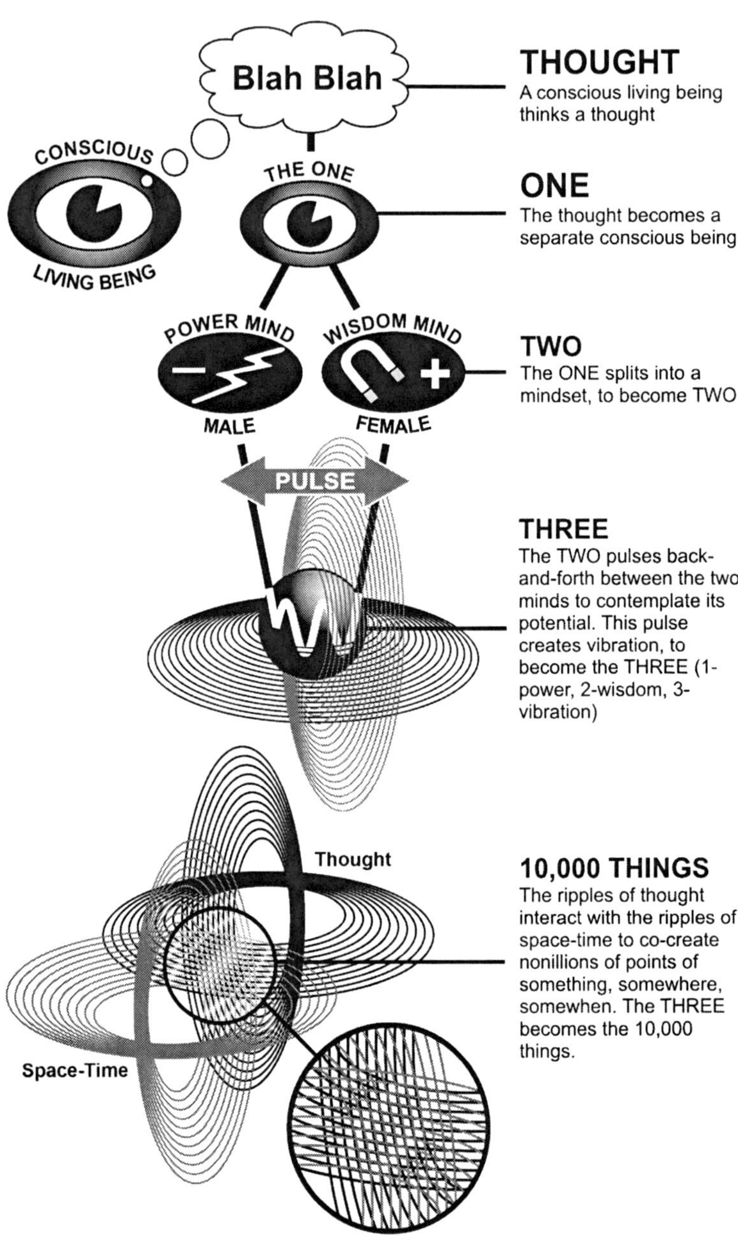

(Figure) The co-creation family tree showing the process of a thought as it follows the co-creative path of ONE-TWO-THREE-10,000-things.

THE 10,000 THINGS

The term '10,000 things' is not literally ten-thousand things but a metaphor for 'many things'. A conscious-living-being co-creates itself, as the thought of itself, because the thought of itself creates the electromagnetic vibration of itself, something, somewhere, somewhen.

That vibrational-reality extends way beyond the physical-world, into the astral-world and the spirit-world and probably further into multidimensional cosmology. Vibrational-reality really is many things.

From a human perspective, the vibrational-reality of the human-being extends from the human body, to the human energy field (aura) to the subconscious mind of the astral-world, to the Soul of the spirit-world. We can only perceive the 'icy' physical body but our vibrational-reality also includes many 'watery and steamy' things we cannot see. There are even points of two-dimensional and one-dimensional reality *below* the vibrational-reality of the physical-world as well as corresponding points of reality in parallel universes. We cannot perceive any of these either.

I pinched that term '10,000 things' from Chinese philosopher Lao Tzu from his now famous quote of the 'Tao (GOD) becomes one, one becomes two, two becomes three and three becomes ten thousand things'. Lao Tzu was referring to (at least in my opinion) the co-creative process of thought. Every thought, from any conscious-living-being (the Gods, Souls, humans), follows the co-creative path of ONE-TWO-THREE-10,000-things.

The vibrations of the Cosmos had no other vibrations to interact with and so there were no 10,000 things - not directly, anyway. The '10,000 things' came later as the *indirect* result of other conscious-living-beings co-creating things as their ripples of thought interacted with the ripples of space-time.

In the case of the First Being, the 10,000 things represent the nonillions of points of double-amplitude, co-created by thought-waves of the First Being's self-aware contemplation interacting with the ripples of space-time. This produced nonillions of points of something, somewhere, somewhen, that extended from the physical-world to the astral-world to the spirit-world - the form, function and phenomena of the First Being. The First Being became the first co-created element of vibrational-reality.

The First Being inherited its potential from its 'parent' the Cosmos. When it contemplated this potential, it was actually contemplating the potential of the Cosmos, since the Cosmos thought the *thought of the first being,* and a thought-wave (ripple) is an expression of potential (potential in motion). When the First Being contemplated that potential, its pulse converted that potential into electromagnetic vibration. That vibration clashed with the ripples of space-time to co-create the vibrational-reality of itself, something, somewhere, somewhen, in the *image and likeness* of its 'parent' the Cosmos. In short, the First Being was just as GODlike as the Cosmos.

THE SECOND BEING AND THIRD AND ...

The First Being then co-created a Second Being in the exact same way - ONE-TWO-THREE-10,000-things. The Second Being then co-created a Third Being and the Third Being then co-created a Fourth Being and so on. How many of these beings co-created other beings? Who knows? One theory is that Lao Tzu's 10,000 things was actually referring to 10,000 Master beings created by the First Being and then those 10,000 master beings created 10,000 more Master beings and so on.

Either way, each new being was co-created with the same amount of potential as its parent, with each parent being created in the image and likeness of the Cosmos. In other words, each new generation of beings was GODlike. These beings were not only direct descendants of the Cosmos but they were all *intimately connected* as layers of powerful and wise potential. These are the spiritual dimensions or layers of the spirit-world. They have also been referred to as 'The Stream', 'Soul-groups', 'The Galactic Federation', 'Abraham' or the Catholic 'Holy Spirit'.

As *separate* beings (even though they are *one*, they often appear to humans as individuals) they represent the Gods in Greek mythology, religious teachings and indigenous stories. In more modern times, we have labelled these Gods as 'Spirit Guides', 'Masters', 'Angels', and my favourite 'the Soul'. Various *physical* reincarnations of the Gods have been called many things throughout history including Buddha, Mohammed and Jesus to name a few.

Being raised a Catholic myself and taught to morbidly fear any person that talks about Gods, especially those weird elephant-looking dudes with more than one set of arms (no disrespect to any Hindus but you have to admit, Viniyaka does look pretty funny), I was a little nervous about using the term 'Gods'. Then I thought, 'Stuff it!'. As part of my post-Catholic rehabilitation program, I will embrace the Gods. The more Gods we have the better, I say.

Humans cannot perceive GOD (it's nothingness). Humans cannot perceive the Cosmos because it's a framework of space-time. Humans *can* perceive the Gods and it's actually the Gods (including our Souls) that have presented themselves to humans over the years, in whatever form best suits that culture, whether that's an elephant or a man with a beard sitting on a cloud (which I'm sure, Hindus find *hilarious*).

If westerners had more elephants, like the GODlike majestic creatures that *used* to roam about India like stray cats, I'm sure our Gods would be more elephant-like. Maybe we should have a God that looks more like Ronald McDonald; 'You want prayers with that?'

The Gods used the same ONE-TWO-THREE-10,000-things process to co-create all the form, function and phenomena of all the stuff *within* the Cosmos. The Gods co-created our galaxies, stars, planets, living-beings and Cosmic phenomena.

The Big Brrr vibration was a separate event that transformed nothingness into the space-time framework of the Cosmos. *Then* the Cosmos co-created the First Being, *then* the First Being co-created other Gods who co-created other Gods. *Then* the Gods co-created all the stuff within the Cosmos. But all of this happened in a few *nonillionths of a second* as part of the infinite expansion, immediately following the Big Brrr. This is why cosmologists think the Big Brrr was a Big Bang. But it wasn't. The Big Brrr was a soft velvety boom of nothingness expanding into a framework of space-time. *Then* the Gods turned on the lights and added stuff! It simply looks like a Big Bang when you combine it all together.

Even more remarkable, the Big Brrr is still happening (expanding) and the Gods are still co-creating our Cosmos, mainly because vibration never stops. A thought is forever. Only the amplitude (reality) fades and only as you move further away from the centre and considering the effect of a thought is infinite, that fade takes a *very long* time. In human terms, a thought is forever. The thoughts of the Gods are still vibrating and that vibration is still co-creating stuff within the Cosmos, including our Sun, our planet Earth and the creatures, plants and geography of this planet. The thoughts of the Gods are still co-creating our human bodies, the tree in your front yard, your cat and even your car! Your thoughts are still co-creating the events in your life now!

SUMMARY

- The Cosmos began with a Big Brrr rather than a Big Bang.
- The Big Brrr only created ripples of space-time that became the Cosmic-pond.
- The Cosmic-pond holds space-time together as form, function and phenomena. It's what makes the illusion seem real.
- The Cosmic-pond is alive, thinks for itself and morphs to suit whatever form, function and phenomena is required, depending on the interacting thoughts of other conscious-living-beings.
- The Cosmic-pond supplies a medium or surface for thoughts to vibrate within.
- The frequency of vibration determines the type of vibrational-reality.

- The amplitude of vibration determines how real the vibrational-reality is.
- The thoughts of conscious-living-beings create ripples in the Cosmic-pond.
- When those ripples interact with the existing space-time ripples of the Cosmic-pond, they create points of double-amplitude loud enough to co-create something, somewhere, somewhen.
- The Cosmos originally had no other beings to think thoughts and so there were no other sources of vibration to interact with the vibrations of space-time and create anything real.
- The Cosmos created the First Being to enable co-creative interaction.
- It co-created the First Being with ONE-TWO-THREE-10,000-things.
- **THOUGHT** - A conscious-living-being *thinks* a thought.
- **ONE** - That thought comes alive and thinks for itself as a separate conscious-living-being. It becomes the ONE.
- **TWO** - That ONE then contemplates its own potential and to avoid its potential cancelling out to zero, it splits into two minds (mindset). The ONE becomes the TWO.
- **THREE** - The TWO then pulses between its two minds to separately embrace its powerful and wise potential. That pulse creates vibration. The TWO becomes THREE (power, wisdom and *vibration*).
- **10,000** - The vibration of the THREE then interacts with the ripples of space-time to co-create nonillions of points of double-amplitude, loud enough to co-create vibrational-reality within space-time, as something, somewhere, somewhen. These points of vibrational-

reality extend from the physical-world to the astral-world to the spirit-world and coalesce into the form, function and phenomena of the 10,000 things. The THREE then becomes 10,000 things.

- The First Being co-created itself, as the thought of itself, because the thought of itself created the electromagnetic vibration of itself, something, somewhere, somewhen - in the image and likeness of the Cosmos.
- The First Being co-created many other conscious-living-beings, using the co-create sequence of ONE-TWO-THREE-10,000-things.
- These beings became the Gods and each God was as GODlike as the Cosmos.
- The Gods then co-created all the stuff within the Cosmos.
- This all happened in a few nonillionths of a second. It all looks like a Big Bang when you combine it all together.

CO-CREATION

All thoughts become self-aware and then go off to do their own thing. They are like little children, scurrying about the astral-world and the spirit-world getting up to all sorts of wonderful mischief (innocently, beautifully, purely and perfectly). So, consider the conscious-living-being (the thinker) to be the 'parent' and its thoughts to be the 'child'. The parent thinks the child, the child comes alive, has its own free will, thinks for itself and then co-creates *itself* as 10,000 things (stuff) that extend from the physical-world to the astral-world to the spirit-world. Every thought from every conscious-living-being, follows this co-creative sequence of ONE-TWO-THREE-10,000-things.

All co-creation is holographic because potential is holographic. What does that mean? We're all very familiar with holograms, the rainbow coloured, three-dimensional images you often see on credit cards or ID cards. Proper three-dimensional holograms display a fascinating phenomenon. If you create a master holographic image that's 800 mm x 600 mm in size, such as the Visa™ dove image and then cut that image into smaller pieces, each and every one of those smaller pieces contains the complete original dove image!

Potential does much the same. No matter what is co-created, no matter how big or small, the child inherits the exact same potential as the parent. It's like Matryoshka dolls, where the smaller doll fits *inside* the larger doll, over and over again.

How does this happen? The child *is* a 'living breathing' thought. A thought is potential in *motion* (vibration) - and so the child is a 'living breathing' manifestation of the parent's potential - or at least the potential behind the thought. When the child contemplates its powerful and wise potential, it's actually contemplating the powerful and wise potential of its parents. As the child pulses back- and-forth between its two minds to contemplate that potential, it transforms that potential into the vibrational-reality - the 10,000 things *of itself*. In other words, the child has co-created *itself* as an exact copy of its parents.

But if every being did this then every being would be the same and there would be no evolution. That's why each child has free will to embrace and express that potential *individually* and hence uniquely, ensuring the co-creation of itself, is slightly *different* from the parent. In other words, the child co-creates itself in the *image and likeness* of its parents - same but *slightly* different.

The vibrational-reality of the child may resemble the adults but is *unique* to the child. This ensures each new generation of 'children' evolves. The points of vibrational-reality are also heavily influenced by the interacting vibrations of other conscious-living-beings (the 'community') and so the vibrational-reality of the child is also collective. The end result is a vibrational-reality that's similar to the parents, unique to the child and influenced by the community. It's co-created.

CO-CREATION IS PASSIVE

The co-creation process is passive because the child co-creates itself in the image and likeness of the parent, *rather* than the parent directly creating the child. In that regard the parent didn't really *directly* create the child even though the parent thought, *the thought of the child*. The parent *indirectly* created the child, because the child created *itself* and in its own individual uniqueness and in conjunction with the community. This is not surprising because everything comes from the source of GOD and GOD does not do (create) anything. That means *everything* does not do anything and so the creative process is always passive.

This was GOD's original plan of 'I shall co-create all things in the image and likeness of me, and observe all things, free, as they become of me.'

GODLIKE CO-CREATION

All conscious-living-beings create this way and I am making a clear distinction between conscious-living-beings and humans. Conscious-living-beings have GODlike powerful and wise potential and they fully embrace that potential, when they think. They're usually spirit-world beings like the Gods, your Soul and the Masters. Humans have limited potential and rarely fully embrace that potential. Not a judgement, just a fact and it's all explained later.

Most Cosmic co-creation is done by the Gods. The Gods all have the same powerful and wise potential as the Cosmos but they still express that potential *individually* as separate beings, ensuring their co-creations are pure, perfect yet *different*. They also co-create within a pure and perfect community of Gods and spirit-beings. The end result is the Gods co-create children that are always pure, perfect but *slightly* different, whether that's a star system, a planet or a cell in your body. The vibrational-reality of their children often extends from the physical-world to the astral-world to the spirit-world as 10,000 things.

The children are also separate conscious-living-beings, and they co-create children of their own. The children inherit the potential of their parents (the Gods), expressing that potential individually within a unique community (spiritual dimension). The end result is again, stuff that's similar to the Cosmos, unique to each child and influenced by a pure and perfect spirit-world community.

It's 10,000 pure and perfect things, co-creating 10,000 pure and perfect things, co-creating 10,000 pure and perfect things … infinitely. The final result is a pure, perfect, yet beautifully diverse Cosmos where everything exists inside everything else as a slightly different version of the larger system preceding it. This is often referred to as the microcosm within the macrocosm and its effects are everywhere - literally.

For example, there are trillions of galaxies, each one perfect, yet none of them are the same. A river system viewed high up from an aircraft looks exactly like the veins in the body, or the veins in a leaf, or lightning. High-speed film footage of traffic stop-starting in the city looks exactly like blood cells stop-starting as they're pumped through veins and arteries. The microscopic grooves inside a vinyl record look like a mountain range. City lights from a distance look like stars in the sky. A flower is perfect, yet each petal is slightly different and the petals are never perfectly symmetrical. Every human fingerprint looks similar but is unique.

It's the same potential, copied into different systems and then expressed differently, influenced by the community surrounding it. That creates different frequencies and amplitudes of vibrational-reality that emanates from what is essentially the same source of potential - GOD. The *difference* can be expressed structurally, functionally or phenomenally.

We humans are already very familiar with holographic co-creation because we do it (literally) with sexual co-creation. Two parents co-create a biological child. The potential of the adults is passed onto the child and then the child expresses that potential individually and hence uniquely, influenced by the community that surrounds it. This explains why a child looks like the parents, yet behaves differently due to its individuality (nature) and its environment (nurture).

There is a curious phenomenon with generational potential. As parents get older they uncover more and more potential, which makes them more 'pure' (goldlike). As each child is born, this ever increasing level of potential-purity is passed onto the child, so that the youngest child often exhibits the most purity.

If you look at a large family, especially where there is a great age difference between the youngest and the eldest child, you may notice that the youngest child often exhibits more purity. They're often purer in nature, purer in health and (dare I say it) better looking than their elder siblings. Their skin, hair, teeth, posture and general appearance is often purer. That generally makes them appear more appealing, more attractive, more beautiful … or however you want to say it so that you offend the least number of older siblings as possible.

The same phenomenon applies to generations as well. The people in those old black and white photos of yesteryear weren't just weird looking because they weren't smiling! They're a weird looking bunch! They have less potential than the current generations and this is physically expressed in their appearance. I mean really, look at their faces. They're not ugly, they are … impure-erer!

(Figure) May 1862. 'Yorktown, Virginia (vicinity). Group before the photographic tent at Camp Winfield Scott.' From photographs of the Peninsular Campaign, May-August 1862. Wet plate glass negative by James F. Gibson.

Cosmic holographic co-creation is far more mind-blowing - more than I can describe because of its multi-dimensional complexity. A better way to demonstrate the complex-beauty of Cosmic holographic co-creation is by viewing a Mandelbrot set. Mandelbrot sets were made famous by the mathematician Benoît Mandelbrot and a mathematically-derived Mandelbrot set creates an infinitely complex image based on an extremely simple equation involving complex numbers. This image shows a colourful pattern, inside a larger identical colourful pattern, inside a larger identical colourful pattern, and so on … infinitely.

The following pictures help demonstrate how things can be similar in shape (same potential) but different in size (different expression of that potential), depending on the community (section of the Mandelbrot set) surrounding it. The white square in each picture shows the zoom area that makes up the next picture.

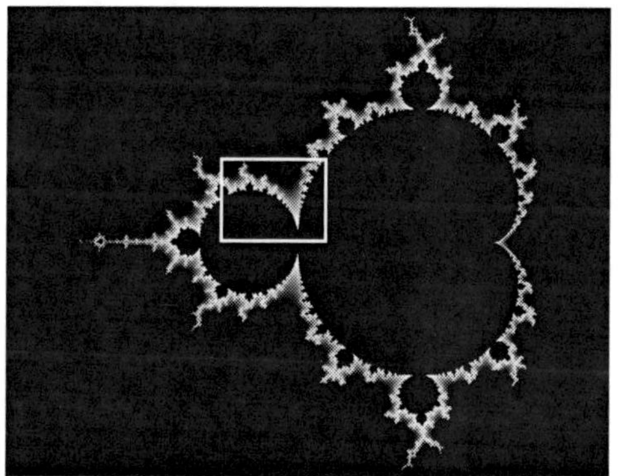

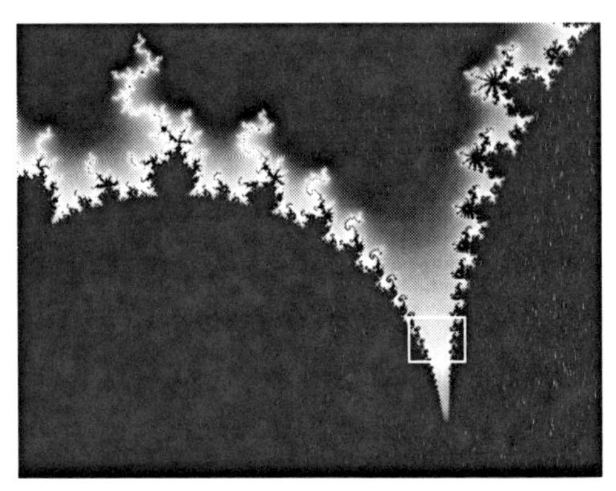

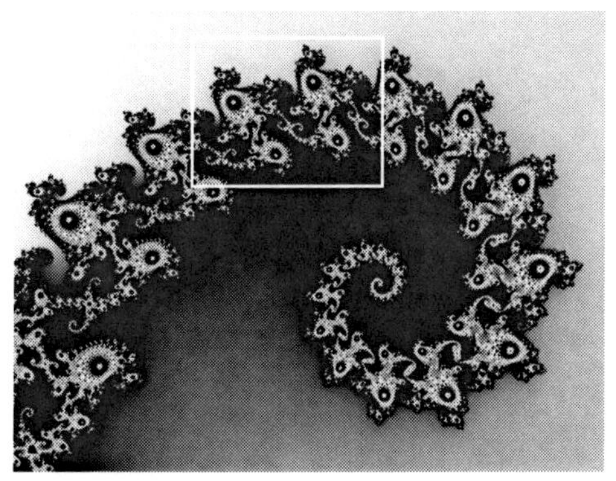

(Figure) Zooming into a Mandelbrot set. Initial image of a zoom sequence with a frame indicating the position of the next image. Created by Wolfgang Beyer, courtesy of Wikimedia Commons.

You can also see these Mandelbrot sets, live in motion at www.yoooge.com. This will give you a fascinating (and meditative) real-time journey into the holographic structure of the Cosmos. Go to the 'Links' section and then click on 'The Mandelbrot Set'.

HUMAN CO-CREATION

As humans we co-create using the exact same ONE-TWO-THREE-10,000-things sequence, except we don't fully embrace our potential. Our thought-waves don't have the amplitude required to co-create stuff *there and then* in front of us, like the Gods can. Firstly, we are born into each lifetime with a predetermined and limited amount of potential. That's part of our evolution. Secondly, very few humans fully embrace their full potential and think with one hundred per cent power and wisdom. The potential is always resisted, diluted or blocked by subconscious bad habits of negative thought.

Our thought-waves create ripples in space-time but they're just not loud enough to directly create vibrational-reality. When the ripples of our thoughts clash with the ripples of *other* human thoughts they create points of double-amplitude. When the ripples of our thoughts interact with thousands of thoughts of millions of people around us, they create nonillions of points of double-amplitude that are often loud enough to create reality. Those points of double-amplitude then interact with the existing space-time ripples of the Cosmic-pond to create nonillions of points of vibrational-reality that stretches from the physical-world to the astral-world to the spirit-world.

What does that vibrational-reality look like? That vibrational-reality is always an event. Why? The rule is simple: If it's co-created by one conscious-living-being, it's a thing. If it's co-created by two or more conscious-living-beings, then it's an event. Instead of us co-creating some*thing*, somewhere, somewhen, we humans co-create some*phenomena*, somewhere, somewhen.

We already do this physically. If we create something on our own, it's often a *thing* such as an email, a piece of artwork or cup of coffee. When we get together as a group we often create *events* such as the Internet, an art exhibition, or a café.

We humans don't have enough potential to instantly co-create things 'out of thin air' like the Gods do, such as planets, stars or solid objects. We can't think of a hamburger and 'poof!' it suddenly appears there and then in front of us as manifested reality. We have to buy the bun, cut the cheese, cook the meat, etc.

And for good reason. Our human fascination with Internet porn, violence, greed and control has pretty much disallowed us the privileges of *instant* creation. Can you imagine the chaos this would create as people, thinking of trees, co-create trees in the middle of the highway during rush hour? Can you imagine the disease this would create as people expressed hate to one another? Can you image what teenagers would do if they had that kind of power?!

We're deliberately limited (self-limited) to the co-creation of events. We can easily co-create the *event* of a hamburger. When we think of a hamburger with strong desire (amplitude), co-created-hamburger-events suddenly appear, all around us. We might open the newspaper or switch on the TV and the first thing we see is a hamburger advertisement, whereas prior to that you've never noticed those ads before. A friend you haven't seen in ages might ring later that day and explain how he's depressed because his wife left him for a hamburger patty salesman. You might go out to lunch and they don't have your favourite meal on the menu that day but they say 'Hey listen, we do a great hamburger?'

These are not coincidences and the way you can tell is that they happen outside of the ordinary routine. In the above examples, you never noticed those ads before, the friend rarely rings and your favourite restaurant always has your dish on the menu. As I always say, 'Spilling a cup of coffee is a coincidence. Spilling a cup of coffee so that it makes an elephant-shaped puddle, the same day you have to make a decision about travelling to India, to ride elephants, is not!'

Another great example of co-created eventual reality is money. Humans don't have the potential (yet) to co-create a million dollars cash out of thin air, but as soon as we *really want* that million dollars, million-dollar-events suddenly surround us. We suddenly get a million dollar idea. We hear people say the word 'million' a 'million squazzillion times'. We (for no reason) glance over to the right at the same time a bus drives past displaying a huge lottery commercial. We watch TV only to find it's a documentary about a very successful person who has just made their first million(s).

It doesn't mean we have co-created a million dollars, yet, but the million-dollar-event is unfolding. The trick is, do we have the awareness, patience and potential to make it happen? Do you?

UNSEEN CO-CREATION

Human co-creation is mostly done on an astral and spiritual level as *unseen* co-creation of vibrational-reality. When you dream of a new car, the vibrational-reality of that car is instantly co-created. That event exists all around you as 10,000 things that extend from the physical-world to the astral-world to the spirit-world, as various different amplitudes and frequencies. That physical-world-reality part of that co-creation is there also but you can't see it, physically. It's vibrating a frequency just outside the range of human perception. It's a child, it's alive and thinking and it's waiting to come into your world. The rest is up to you.

This vibrational-reality is still attached to the vibrations of your mindset and morphs to suit whatever you're thinking. Your thoughts also attract, repulse, switch on and switch off the physical-world part of that reality. The end result is that vibrational-reality comes and goes as your thoughts constantly change.

If you maintain a joyful, passive, expectant focus on your new car, the vibrational-reality of that car will get louder and louder. It will eventually manifest as a real, physical-world, new-car-event. In that regard we're *eventual* co-creators. We co-crate *event*-ually (eventually). That classic saying, 'Shit Happens!' is truer than most people think.

Every single thought you think does this. We *think* we're just thinking stuff but each of us is a 'parent' and every single thought we think co-creates a 'child'. That child scampers off into the Cosmos to produce the vibrational-reality of 10,000 things that extends from the physical-world to the astral-world and to the spirit-world.

You might be thinking, '%$#@, I think about stuff *all the time*. That's a lot of children running around in the astral-world and the spirit-world, creating a whole bunch of vibrational-reality!' I say, 'Yep!' Why do you think the

Cosmos is expanding at 75 km per second per 3.26 million years!

THE YOUNIVERSE

Even though the vibrational-reality is co-created by the interaction of nonillions of other ripples, from the thought-waves of millions of people thinking thousands of thoughts; the vibrational-reality is more real to the thinker because the amplitude of the thought is strongest at the source. In that regard, we are all the centre of our own private universe, or *you*niverse as I call it.

Just imagine, seven billion people, living within seven billion completely separate, yet interactive *you*niverses. That's very possible since everything comes out of nothingness and returns to nothingness. Thought-waves can create ripples, within ripples, within ripples, in the same way that music can contain the sound waves of many different instruments combined.

That vibrational-reality can manifest into our world as any sort of event such as luck, support, opportunities, ideas, accidents or even violence. The type of event is up to the individual because **every event we co-create in our lives perfectly matches the vibration of our thoughts.**

Every single thought we think becomes 10,000 things and those 10,000 things *always* perfectly match the vibration of our thought. Every one of those 10,000 things surrounds your *you*niverse from the physical-world to the astral-world to the spirit-world. We are literally surrounded by nonillions of events good or bad. We can't see them but they're there.

The magnetic-potential within our thoughts will *attract or repel* one of those events (it will come or it won't - both are events) *into* our *you*niverse, in a way that matches the vibration of what we think about most. The electrical-potential in our thought-waves will then switch one of those events *on or off* (it will happen or it won't) in a way that matches the vibration of what we think about most. Those events are then facilitated and delivered (co-created) by the community surrounding us, in a way that matches the vibration of what we think about most. We get what we 'asked' for - good or bad.

For example, people who worry about being burgled are often burgled. Optimistic people always seem to get all the opportunities. Angry people attract events that make them angry. People who are truly grateful for everything always seem to have an abundance of everything. People who complain about money never seem to have enough money.

It's no coincidence. It's Cosmic law. Nothing is by chance, there are no accidents and there are no victims. *Nothing* can enter your *you*niverse unless it matches (similar but different) the vibration of what you think about most because what you think about most is co-creating, attracting, repelling, activating and deactivating the vibrational-reality of your own private *you*niverse - no exception.

Your eventful life, *now*, is a direct reflection of what you think about most (remembering that up to ninety percent of thought can be subconscious). Events are only a 'coincidence' or a 'miracle' to those who *don't* understand the co-creative process of ONE-TWO-THREE-10,000-things.

EVENTS WITHIN EVENTS

Just as Cosmic co-creation is holographic, human co-creation is also holographic. Events exist inside other similar events. For example, have you ever noticed that when you really *want* a new car, you suddenly see that model of car everywhere? It's no coincidence and it's not *just* because you're suddenly aware of that model. It's the vibrational-reality of the smaller event (other people driving that model of car) existing inside your larger event (you owning that car) as a microcosm within the macrocosm.

Similarly, have you ever noticed that when you forgive someone, someone forgives you? Have you ever noticed that if you think of a friend that you haven't seen in a long time, they suddenly ring? Have you ever noticed that things happen in threes - big, bigger, biggest? Have you noticed that the majority of the people who win lotto, are already financially OK (they have wealth, they think wealth and so they create more wealth)? All these events are simply smaller, similar yet slightly different, community influenced versions, of the larger event preceding it.

Most people brush this idea aside like a wasp because they find it hard to believe that we humble humans could have that much power. But what these people don't understand is that human co-creation is not individual, it's super-collective. The best way to illustrate this is with music.

Music is a complex mix of vibrational frequencies and amplitudes that creates ripples in the air via a pulsing speaker, which our ears translate into sound. Thought is a complex mix of vibrational frequencies and amplitudes that creates ripples in space-time from a pulsing mindset, which our senses translate into eventual-reality. The processes are both the same and the effects are very similar. That's *why* humans have been so tantalised by music and that's why music makes us feel so good.

The vibrations of music are complex. For example, if you pluck a guitar string, that string rapidly moves back-and-forth, creating a vibration in the air (soundwave) that we experience as the sound 'plongggg'. That's just one string on one instrument. Now imagine a grand piano that has two hundred and thirty 'guitar strings' within it and imagine a musician playing a complicated, fast, piano concerto. Can you imagine the complexity of the ripples produced by that grand piano?

Now imagine the rest of the orchestra, with over eighty members, all plucking, playing, bowing, blowing, hitting, and singing complicated mixes of vibrational frequencies and amplitudes. The complexity of all those vibrations interacting, resonating and over-toning together as music is *staggering*.

Then imagine the complexity of nonillions of your thought-waves interacting with the nonillions of vibrations of millions of people, thinking thousands of thoughts. That's loud! Then combine that with the nonillions of vibrations created by the contemplation of plants, animals, Mother Earth and the Gods. That's even louder. Now mix all that together with the loudest amplitude of them all - the existing space-time ripples of the Cosmic-pond. There is enough amplitude there to *easily* co-create nonillions of points of vibrational-reality as the form, function and phenomena of some*phenomenon,* somewhere, somewhen.

The creative power is super-collective. So yes, as individuals we don't have that much power. As a group we can create miracles. 'When two or more people are gathered together in my name …'

SCIENCE WILL NEVER BELIEVE THIS

For the rational scientists, this theory has to be proven in a laboratory before they can believe it. That's going to be tricky, considering that scientific observation influences the outcome of their own observation.

Rational scientists will *always* be able to prove that something is nonsensical whoopy foof because science has a natural bias towards rationality. Its guffawing disdain for metaphysics creates biased thought-waves that collide and interact with all the other biased scientific thought-waves, co-creating a vibrational-reality of nonsensical whoopy foofness. It's the scientific equivalent of peeing in your own pond and then complaining about the smell!

Speaking of which, scientists believe that reality is made from tiny vibrational strings of energy. This is the basis of String Theory. I believe these strings are nothing more than the ripples in the Cosmic-pond. They're not strings, they're rings although I hesitate to use the term 'Ring Theory' (Australians will understand this humour, more so than others!).

As we pulse back-and-forth between our minds, between the two planes of space-time, yet moving *forward through time,* there is an electromagnetic twisting force within that pulse. This twisting force bends these rings into twisted loops that look remarkably like infinity symbols (interesting!). You get the same effect if you hold a ring of wire with each hand and then twist one hand clockwise and the other hand anti-clockwise. Maybe we could use the term, 'Loop Theory', or maybe even 'Loopy Theory'.

The reason why strings are so small is because their frequency is in the nonillion hertz range - small frequencies equals small waves, which equals small things. The reason why they form the basis of everything is because everything is made of rings (ripples in the Cosmic-pond).

The reason why scientists haven't been able to fully explain string theory is because in the process of moving away from their own pond, because they peed in it and it smells, they peed all over metaphysical theory instead. Yet if they had only been open minded enough to consider the 'other cause of things', they'd have all the answers they need.

What I find amusing is that metaphysicians (or at least the ones I know) study science to help form a more balanced point of view. Yet few scientists (or at least the ones I have met) will even consider metaphysical theory. Maybe I just need to get out more. Maybe I am just co-creating points of vibrational-reality that manifest as closed-minded scientists? Mmm!

SUMMARY

- Every thought from every conscious-living-being follows this co-creative sequence of ONE-TWO-THREE-10,000-things.
- The thinker is the parent. The thought is the child.
- Each child co-creates itself in the image and likeness of its parents with the exact same potential.

- Each child has free will to express that potential *individually* and hence differently.
- The end result is a vibrational-reality that's similar to the parents, unique to the child and influenced by the community.
- The Gods co-create children that are always perfect and pure. Those children then co-create children of their own that are always perfect and pure.
- The co-creation is always holographic - similar but different.
- The end result is a beautifully diverse Cosmos where everything exists inside everything else as a slightly different version of the larger system preceding it.
- Humans don't have enough potential to create things but they can co-create events.
- A God creates some*thing*, somewhere, somewhen. Humans co-create s*ome phenomena,* somewhere, somewhen.
- Most human c*o-creation i*s unseen. It exists all around us as the 10,000 things that extend from the physical-world to the astral-world to the spirit-world. It only becomes physical-world-reality when we attract and activate that event into our *you*niverse, with our thought-patterns.
- Every event we co-create in our lives perfectly matches the vibration of our thoughts.
- Nothing is by chance, there are no coincidences, there are no accidents and there are no victims.
- *Nothing* can enter your *you*niverse unless it matches (similar but different) the vibration of what you think about most because what you think about most is co-creating, attracting, repelling, activating and

deactivating the vibrational-reality of your own private *you*niverse - no exception.

- The vibrational-reality of the smaller events exist inside larger events as a microcosm within the macrocosm.
- Scientists peed in their own pond!

THE MYTH OF GOD

That's the Cosmos. Now let's talk about god. Here's where this book gets a little bit funky and when I say funky I mean, 'Oh my, I have just stepped in a dead whale!' kind of funky. Are you ready …? There is no such thing as god?

OK, I said it, there it is, it's out in the open and now I feel a tremendous amount of relief. There is no such thing as god. There is no such thing as god. There is NO such thing as god. Weee! Oh it's so liberating.

And just to be perfectly clear, I'm talking about the *concept* of god; the god that people *think* is god; the lower case version of 'god '; the bearded deity that sits on a cloud and punishes people for their sins; the god described in the Bible; the Christian god that has churches built in his name and is worshipped by millions of Christians, all over the world. *That* idea of god is so fundamentally flawed, so 'human', so not-GOD, that it's bordering on the ridiculous. In that regard, there is no such thing as *that* god (I will explain in detail).

The statement, 'There is no such thing as god' would have gotten the back of your legs slapped by a nun in the Catholic Primary School yard - for sure! She would have held you aloft with one crusty, nun hand so she could get a good go at the slapping with the other crusty, nun hand as she enunciated every syllable of your naughtiness with each slap. You (slap) are (slap) a (slap) ver (slap) ree (slap) norr (slap) tee (slap) boy (slap) … Yes this happened to me when I was six years of age at primary school, although I wasn't denouncing GOD's singularity. I got slapped because I stole a doughnut from Tania Smith. It had jam in it and was worth it.

Let's face some metaphysical facts. OK, there is no such thing as a metaphysical fact, so let's face some metaphysical *theory*. Firstly and most importantly, everything is *one*, especially within the Cosmos, and so it's all GOD. Although if you want to be picky, GOD is nothingness and everything else is actually not-GOD. But hey … who's going to audit that? Now let's line up all the versions of what could be *considered* to be a GOD, or God or god. To make it simple, I call these God-Candidates:

1. **God-Candidate 1 - GOD**. This is the real GOD, the original GOD that existed before creation. She's the GOD that thought 'I am that I am' to explore Her own Divine Magnificence
2. **God-Candidate 2 - The Cosmos**. This is the thought 'I am that I am' that came alive to become the ONE, split into the TWO and then became the space-time vibration of the THREE, which is now the Cosmic-pond.
3. **God-Candidate 3 - The First Being**. This is the First Being, co-created by the Cosmos as the first point of reality.
4. **God-Candidate 4 - The Gods**. Then there is a whole bunch of other beings co-created by the First Being. They're *all* Gods and equally GODlike as each other because each God inherited the same level of potential as the First Being and so are equally *as* GODlike.

5. **God-Candidate 5 - god.** This is the god that most people think is the real GOD, commonly known as 'God', the dude with a beard sitting on a cloud that's described in the Bible. It gets a lower case 'g' because that idea of god is essentially flawed and so does not deserve a capital 'g', in my book (I actually have a book and so I can say that!).

Confusing isn't it? More to the point, who in the hell have we been *worshipping* all these years? So let's go through each God-Candidate one by one to help clarify the true concept of GOD, God, the Gods and god. Let's start with GOD.

GOD

No human has ever had a direct experience of GOD because GOD is nothingness and nothingness cannot be comprehended, imagined or perceived by humans. It's pure potential without vibration and we can only perceive that which vibrates. That immediately wipes out any chance of any human *ever* having a direct experience of GOD.

Nothingness is stillness (still, passive, static). Anything that's not stillness must then be not-GOD. In the Cosmos, everything, everywhere and everywhen is vibration. Vibration is the exact opposite of stillness. It's always buzzing, active and changing. If everything, everywhere and everywhen within the Cosmos is vibration then everything, everywhere and everywhen within the Cosmos must be not-GOD. We humans live in the Cosmos and so there is no possible way we can perceive that which is *not part* of the Cosmos. Simple!

THE COSMOS

The Cosmos is a dark, empty and hidden framework of space-time. That immediately wipes out any chance of any human having a *direct* experience of the Cosmos - unless you die of course where you get to cross the pond (explained in the 'Death' chapter).

THE FIRST BEING

The First Being is the God of the Gods, the Creator of conscious-living-beings; the first 'physical' manifestation of powerful and wise potential. The First Being is a big topic, a whole other book, and I haven't received any more downloads about it so I really don't know. The First Being could be hidden from all of us. The First Being might be Jesus or Buddha. Who knows, it's all connected and all pure and perfect so the frequencies of vibration are very similar.

What *feels* right to me is that although the vibrational frequency of the First Being is very similar to the Gods, its amplitude is too soft to be directly perceived by humans. As a result, it's unlikely that the average person can directly perceive this energy, otherwise we would have heard more about it by now on YouTube™. I would assume that no human can have a *direct* experience of the First Being.

THE GODS

The Gods on the other hand are loud and proud and interact with us humans *all the time*. I love the Gods. I think everything that humankind has ever worshipped as a deity has actually been the Gods, all along. Why? Because that's their job. The Gods are the messengers of Light. They are the bridge-builders between the physical-world, the astral-world and the spirit-world. They are the Spirit Guides, the Masters, and the Angels. They are your Soul. They form part of your subconscious mind. They power your intuition. They are you!

As far as human interaction goes, the Gods have a vibrational energy with a frequency and amplitude specifically tuned to a level that can be received by *all* humans. In a state of joy, *any* human can communicate with their Soul. Consequently the spiritual experience comes from the Soul - at least that's my opinion. It's the source of Divine inspiration, intuition, revelation and enlightenment. I know that's where I get my downloads and I have felt the Divine inspiration of my Soul many times and it certainly feels like what god 'should' feel like. It's overwhelmingly powerful, supremely wise, can be tearfully joyful and is always humble and passive.

And your Soul is the bigger part of you. It's YOU (as in the bigger you, which is YOU). When your Soul communicates with you, it feels like a big brother or big sister has just given you a Divine tap on the shoulder to let you know they're helping you, guiding you or looking out for you. Yet deep down you know it's YOU. Sometimes this feels like intuition, other times great ideas come into your mind, sometimes it's moments of Divine Grace and often it's opportune events (Souls like to work together in groups).

What kind of communication you get from your Soul depends on your level of joy. The main reason (if not the only reason) artists and musicians get inspired by their muse is because whilst they're creating art or music they're often in a relaxed state of joy. This state opens the channel for Soulful communication - downloads. Their muse is actually their Soul.

The same applies to meditation, chanting or prayer. All three practices involve *less* thought. Less thought means less negativity, which means more joy, which means a more open channel of communication with the Soul. When the channel's open, the Soulful communication flows. You can get the same effect by directly asking your Soul a question, when you're in a really good mood and then quietly waiting for an answer.

I can see why people think this is god. When you pray to god, your Soul gets the message and then co-creates the 'answer's to your prayers' as events that surround your *you*niverse. If you keep praying and hoping you co-create a *you*niverse that lets those events in so they eventually become reality. That would seem like god answered your prayers, even though it's your Soul.

It makes sense! GOD doesn't do anything, so it can't be GOD. The Cosmos and the First Being are equally passive and observational, so it can't be them. That just leaves YOU - your Soul.

The religious experience is Soul-powered. The spiritual experience is Soul-powered. The feeling of joy that converts normal sane people into born-again Christians, comes from the Soul. "Witch Doctors" and 'Medicine Men' have channelled healing energy from their Soul. People such as Paramahansa Yogananda, The Dalai Lama, Mahatma Gandhi and Mother Teresa were directly guided by the wisdom and power of their Souls. It's not some guardian

Angel from Heaven. It's your Soul. It's not the Holy Spirit. It's your Soul. It's YOU.

Why has this idea not been revealed in mainstream religion? It's such a beautiful concept. It's so comforting. It's so empowering. It's so revelational. The Church deliberately suppressed this idea centuries ago and the reason is simple. If YOU have all this Divine power and wisdom, and all *you* have to do to access this power and wisdom is feel joy, then why would you need a Church? If you didn't need a Church then why would you support the Church? Why would you donate money, land and power to the Church? Why would you obey the "word of god", re-written by the Church to subjugate and control the masses in a way deemed appropriate by the Church?

It's important to note that I am referring to "the Church" as a global organisation, not a personal priest-level or local-Church-community level. We all know that the local Church, its priests and volunteers have done amazing work for millions of people. I'm referring to "the Church's" global campaign of religious domination, which has not always been in the best interests of the people because of its strong ties with power, money, politics and control. That Church has a history of disliking anything that gives people secure control of their own destiny. Even more nun-slappy is the idea that god has a GOD.

GOD (LOWER CASE GOD)

So all this time, Christians have been worshipping a concept of god that is fundamentally flawed. In that regard, that god does not exist, mainly because the concept of that god couldn't be further from the truth. The god that most

Christians or Catholics believe is god is external from them, often angry or vengeful, who chooses and meddles, and who will punish them if they do not do what *he* says.

The actual god, which is their Soul … *is* them, or THEM; it's only way is power, wisdom and joy; it's passive, and unconditional; and its *only* task is to help and guide them through life, so *they* achieve the enlightenment needed to think with more and more power and wisdom, to feel more joy and hence *be* more like their Soul, the real THEM.

Anything other than the above has *absolutely nothing* to do with the Soul and is self-inflicted human misery - negativity. Simple!

Instead of praying to god, who does not exist, pray to your Soul - YOU. In a state of joy, ask your Soul for the inspiration, guidance, answers, ideas or help you need to find more power, wisdom and joy within yourself … and then let go. No need to go to church, kneel down or avoid meat on Fridays. You can 'pray' at home, lazing on the couch, eating pork ribs.

Your Soul doesn't do anything *for you.* It guides and assists you so that you *do it yourself.* Unlike the common misconception that god does stuff for you, you actually have to get off your arse and challenge yourself enough to uncover the potential *within you,* so that you can feel more joy (explained later). All your Soul does is surround your *you*niverse with events to assist you in this endeavour.

THE MAN MYTH OF GOD

It's no wonder religion is so confusing. Take the man-myth of god for example. There are two aspects to the Cosmos -

power and wisdom. Men have always been traditionally aligned with power because it's more of a male-mind-energy. This makes men more aligned with the physical-world. That's *why* we have a world dominated by men. On the other side of extreme-duality, women have always been more traditionally aligned with wisdom because it's more of a female-mind-energy. This makes women more aligned with the spirit-world. This may help explain women's intuition and why most famous psychics are women.

You only have to look at how the various potentials influence certain characteristics, to clearly see how men and women have aligned with each mindset:

POWER - MEN

- Security (adventure, courage, materialism).
- Capability (skill, success, willpower).

WISDOM - WOMEN

- Expression (emotional expression, social interaction, personality).
- Knowing (compassion, intuition, family).

It's not that men are better than women or women are better than men, it's just that they both have their different potential strengths and this completes the duality of the sexes. Men have more potential for the characteristics of power and women have more potential for the characteristics of wisdom.

But that's what started the whole man-myth of god. Men are stronger in power and so tend to be weaker in wisdom. Power without wisdom makes a man weak and a weak man's beliefs can easily be distorted into insecurity, greed, lust, control and domination. History has shown that to be true.

Spiritual teachers and leaders have mostly been men with a historically patriarchal stronghold over education and religion. That's the side-effect of control and domination.

These same distorted men then distorted the image of god by portraying him as a power-hungry man-god who was angry, vengeful and nasty and who would 'burn the shit out of you in Hell' if you didn't do what you were told. They then assumed god must have a church of male leaders who must create laws of punishment (insecurity); grab money, land and jewels from worshippers (greed); enforce celibacy (sexual repression); then rule the world as The Church (control and domination).

They then banned women from playing any role in The Church because they *weren't men* and so could not represent god. Woop, woop, woop (sarcasm alert!). Let's face it, the last thing you want in The Church is a priest with powerful communication skills, great compassion and a higher awareness of spirituality, not to forget a woman's better understanding of love, healing and family!

These same religious men, perceived the Gods, assumed it was god, assumed it was man, assumed it was angry, assumed it was separate from man and so beyond man, and so put a wise angry bearded man on a cloud and called him 'god'.

God has been worshipped ever since as 'Him', 'He' and 'Lord', 'King of the Heavens', 'The Father' or whatever other manly-man label mankind could add to the man-god.

It almost surprises me that god didn't end up being called 'The Great, Bearded, V8 Driving, Beer Drinking, Golf Playing, I'll Be In My Garden Shed Playing With Tools and Gadgets, god'.

THE GIRLY GOD

Metaphysically speaking, the god that religious menfolk have been so proudly tooting their spiritual hooters about is a 'girl god'. I love the irony of that.

Religious menfolk perceived the Gods but assumed it was god. Yet the energy of the Gods extends from the physical-world to the astral-world to the spirit-world. We just can't see them in the physical-world.

Religious menfolk assumed that god lived in the spirit-world. The spirit-world is the female mind of the Cosmos so any deity that lived *in there,* would be a female-mind energy. If you just focus on the spirit-world aspect of a God, then you are only focussing on the female aspect of that God and disregarding the male aspect. It's an unbalanced concept of a 'girl God'.

Religious menfolk did exactly that. They simply ignored the physical-world because it didn't seem Divine enough. They then focussed on the spirit-world. They perceived the Gods, assumed it was god, called it a male god, and worshipped it - even though it was just the 'girl God' aspect. That's hilarious.

You can plainly see the effects (not so hilarious) of an institution that has focussed its attention on the wisdom of the spirit-world (prayer and spiritual teachings), yet dismissed the power of the physical-world (security, sexuality and personal-power). The Church has to take responsibility for its propagation of fear (lack of security), suppressed sexuality (lack of sexuality) and control (lack of capability caused by 'taking away' a person's personal power - the ability to make a decision for yourself based on your own intuition rather than the rules of the Church or god).

Meanwhile, other more spiritually savvy, indigenous cultures - deemed 'savages' by the Church - easily perceived and identified the male and female aspects of the Gods. They perceived the female energy of the Gods that resides within the spirit-world, assumed it was all connected as *one* and so called it the Goddess, Great Mother, or the Sacred Divine Feminine. They perceived the male energy of the Gods that resides within the physical-world, assumed it was all connected as *one* and so called it the God, 'Great Grandfather', 'The Sacred Divine Masculine', or 'The Central Sun'. This created the beautifully balanced concept of a God/Goddess.

What's really interesting, especially if you're a recovering Catholic (Cathoholic) like me, is that some believe the original Lord's Prayer (the 'Our Father') acknowledged a God and Goddess! Many of the original scriptures that now make up the modern Bible were translated from Aramaic into Greek and then into Latin, then old English and finally into modern English, creating one of the worst cases of 'Chinese whispers' in religious literature. Some people believe that the original Lord's Prayer, in Aramaic (the language spoken at the time of Jesus) didn't begin with *'Our Father who art in Heaven ...'* but originally began with *'O Birther! Father-Mother of the Cosmos ...'* *

*Translation by Neil Douglas-Klotz in the book, *Prayers of the Cosmos.*

THE MOON - THE LUNATIC

Historically speaking, religious men have assumed that god exists in the spirit-world and have bypassed the male-God-energy of the physical-world. To help balance this misconception, the Gods co-created the Earth's Moon.

Men relate sunlight to the physical-world. Men relate the darkness of night with the spirit-world and hence god. The Moon is a reflection of the Sun in darkness. The Moon then becomes a subtle energetic and symbolic reminder that within god (darkness of night) there is a physical-world element (sunlight) and so god must somehow exist within the physical-world. This got man to ponder the concept of God/Goddess. The indigenous people got the symbolism, without texting each other or using Facebook™!

The balanced concept of God/Goddess was formed to help humans uncover potential. We get our own potential from within these worlds and so to recognise and honour the creative power, life-force and beauty of Nature (God) along with the wisdom and wonder of the spirit (Goddess) is to recognise, honour and hence re-member that same potential within ourselves. We cannot perceive what we do not have within us.

LUNATICS

Moonlight has a very powerful healing energy that balances the Darkness within weak men. The word 'Darkness' gets a capital 'D' since Darkness is the opposite of Light, which is God-Light. This Darkness is the aforementioned fear, greed, lust, control and domination, which ironically is the same Darkness that prevents men from assimilating a more balanced view of God/Goddess.

The word 'lunatic' comes from the Latin word 'lunaticus', which means 'moon-struck' or 'moon-sick'. It represents the traditional link made in folklore between madness and the phases of the Moon ('Luna'). Our modern version of that word is 'loony' or the 'Loony Bin'.

Lunatics abound on a full Moon because the feminine energy of the full Moon helps balance the Darkness within unbalanced men. When you heal unbalanced men, the *Light interrupts the Darkness* and the Darkness leaves, usually in the form of mental and emotional discord. In men, severe mental and emotional discord is often expressed as anger, violence, depression or suicide. This is why hospital wards all over the world have a *tenfold* increase in emergency ward admissions on a full Moon, the majority of which are men, or their victims.

Most men view the Moon to be just a little bit spooky and consequently created a whole lot of manly folklore about werewolves and lunatics. Women find the Moon to be passive soothing energy. It reminds them that the feminine Goddess is always watching over them, even in Darkness. Consequently, women were more inclined to run around nude in the moonlight to honour the Goddess. This got them into trouble.

THE GODDESS BOND - THE WITCH

Women have historically shared a special bond with the Goddess and over the centuries they have tapped into the female potential of the spirit-world. The increased wisdom gave them better intuitive access to Cosmic knowledge. This helped them become great herbalists, healers and diviners with a broader understanding of human nature and human energy. Dancing in the moonlight only made this ability stronger.

The Church, being filled with fearful-religious-menfolk, misinterpreted these female, Moon-worshipping, herbal healers, as evil. They also didn't want women to have a dominant relationship with god because that threatened their power-hungry, controlling, angry relationship with god. Consequently, they conjured up great stories of hideous hags brewing potions (herbalism), performing spells (healing) and talking with the 'dead' (channelling their Soul). They called them witches and then punished them.

Contrary to popular belief, witches weren't all burnt at the stake. It was simply too time consuming and expensive. Some were hanged or strangled then sometimes the body was incinerated, others tortured and released, but most were publicly shamed before being acquitted and set free (hopefully to pass on the message that if you do witchy-type things, then the Church will do terrible, hurtful things to you). More recently, the Church has simply banned women from being priests.

SUMMARY

- There is no god.
- There is GOD, the Cosmos, the First Being, the Gods and god.
- GOD, the Cosmos and the First Being cannot be perceived by humans.

- The Gods exist at a frequency and amplitude that *can* be perceived by humans.
- Everything that humankind has ever worshipped as a deity has actually been the Gods.
- Your Soul is one of the Gods.
- Your Soul is YOU.
- In a state of joy, any human can communicate with their Soul – downloads.
- The spiritual experience is Soul-powered.
- Distorted religious menfolk, perceived the Gods, assumed it was god, assumed it was man, assumed it was angry, assumed it was separate from man and so beyond man, and so put a wise angry bearded man on a cloud and called him 'god'.
- That god is actually a girly God – hilarious.
- The Moon helps reminds man to ponder the more balanced concept of God/Goddess.
- The Church, as an institution, does not seem to like women very much.

THE AGREEMENT

You are *not* who you think you are. The *you* that *you think is you,* the person reading this book, *is not* the real you. It's only a *very tiny part* of you.

If you step outside the human body for a moment you will notice something amazing. The vibrational-reality of your human-self extends from your human body, to your body's energy field (the aura), to the astral plane (subconscious mind) and then to the Light of the spirit-world.

If you look into the Light of the spirit-world you will notice a single point of Light *within* that Light. If you zoom into that point of Light, you will notice that that point of Light is your Soul. This is the real you, your Soul, YOU.

If you look very closely along that energy line from you to your Soul, you will notice that there is *absolutely no separation* whatsoever between you and your Soul – none! It's all joined, it's all one and it's all just different frequencies of the same vibrational-reality.

'OK, so I'm my Soul and I'm all *one* with it and it's real groovy, blah blah, blah', I hear you say. That's the common reaction by most people to that statement only because most people don't really *get* the significance of that statement. So let me make it bigger for you. Your Soul is one of the Gods. This means the real you, is a God. You are a God:

You

are

a

God!

YOU are one of the Gods co-created by the First Being. YOU are Divine. YOU are a deity. YOU helped co-create the splendour of the Cosmos. YOU are so powerful and wise that YOU are worshipped by many of the other spiritual beings in the Cosmos. That's who YOU really are.

YOU are a spiritual being that lives in the spirit-world. YOU have always lived there - infinitely. YOUR 'body' is made of Light. It has no physical form and extends infinitely as *part* of the Light, although concentrates as a *point* of Light.

YOUR community consists of the Cosmos, the First Being and the Gods. That community is all connected as *one,* and YOU are an integral member of that community. But YOU are also very special and not just because YOU are God but because you have incredible power and wisdom - enough power and wisdom to be specifically chosen to help GOD experience Her own Divine Magnificence.

How? You made an agreement with GOD. That agreement was quite extensive and I will describe it step by step in this chapter. In this agreement, all the uppercase versions of 'YOU', 'YOURSELF' and 'ME' represents YOU (your SOUL), whereas all the lower case versions represent you (the human).

YOU AGREED TO CREATE IMPURITY

The essence of this agreement is that YOU agreed to help GOD experience Her own Divine Magnificence by creating impurity, as a human! I don't use the word 'impurity' to imply judgement of humans, or to expose human flaws. I use the word 'impurity' to help *define* a person's level of gold. Everyone is goldlike in some way. We only focus on the gold in people. But that level of goldlikeness is defined by how impure you are. In that regard, knowing your impurity means knowing your goldlikeness.

Sticking with the original gold example for a moment, GOD is 100% pure gold and that does not exist anywhere in the Cosmos. The Cosmos is not-GOD, which is the not 100% pure gold. It's the illusion of gold, or goldlike and the illusion of gold can extend from Chuk Kam (99.99% pure gold), to 24 carat gold, 22 carat gold, 18 carat gold, 14 carat gold and 9 carat gold (not strictly because there are infinite combinations of purity/impurity but the gold example helps simplify matters).

Carat Purity	Gold Content
Chuk Kam	99.99%
24 carat gold	99.0%
22 carat gold	91.6%
18 carat gold	75.0%
14 carat gold	58.5%
9 carat gold	37.5%

(Table) The impurity of gold.

As a God, before you were even born, YOU knew that everything *within* the Cosmos was initially created pure. Not 100% pure (because that's GOD) but 99.99% pure Chuk Kam. YOU knew the First Being was Chuk Kam. YOU knew all the other Gods were Chuk Kam. YOU knew that everything the Gods created was Chuk Kam. That meant that everything, everywhere and everywhen was 99.99%

pure. That's *very* goldlike and very close to one hundred per cent 100% pure gold. That meant that everything, everywhere and everywhen resembled GOD, *very* closely.

GOD already knew what GOD was. GOD 'wanted' a more expanded and complete view of Her own Divine Magnificence. Consequently, YOU knew that *someone* had to create *something* that was not 99.99% pure, in order to spread the illusion out from Chuk Kam to 9 carat gold. This would then present a more expanded viewpoint of goldlikeness.

From a human perspective, the only reason we understand and appreciate the value and beauty of 24 carat gold is because we've all experienced the cheaper, 9 carat (often gold plated) jewellery. It's nice but … it's not solid 24 carat gold. Otherwise, how would we know? We know how it feels to own a 9 carat gold piece of jewellery. It makes you feel OK. 14 carat gold makes you feel good; 18 carat gold makes you feel special. But solid 24 carat gold makes you feel *very* special. It's heavy, luscious, golden-orange, loveliness. Chuk Kam makes you feel like royalty.

That's an expanded view of goldlikeness. It's all gold*like* (the illusion of pure 100% gold) but it's only from the expanded viewpoint of goldlikeness (9 carat to Chuk Kam) that we can fully appreciate Chuk Kam.

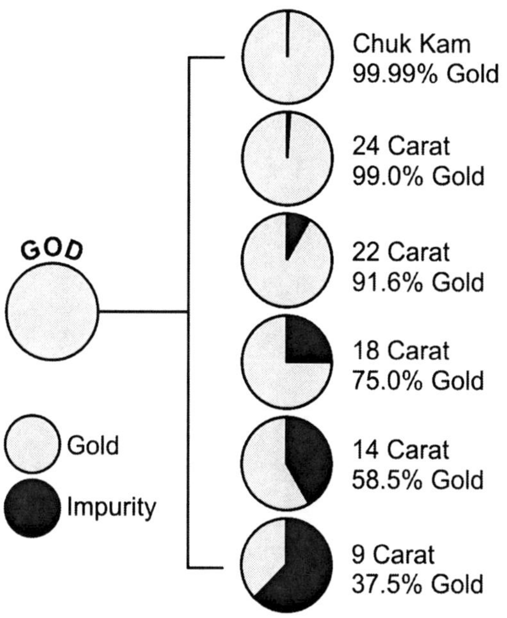

(Figure) The expanded viewpoint of GOD's Divine Magnificence, as if it were compared to gold.

The Cosmic version of goldlikeness is GODlikeness. The more impure a person is, the *less* GODlike they are. The more pure a person is, the more GODlike they are: the greater the variations of impurity, the greater the variations of purity. Given enough time (infinity), every possible version of purity could be created, presenting every possible viewpoint of *GODlikeness* to GOD - an expanded and complete view of GOD's Divine Magnificence.

YOU AGREED TO THE HUMAN EGO

Consequently, YOU knew that the only way to create an expanded viewpoint of GODlikeness was to add some impurity into the mix. The only *place* YOU could create impurity was the physical-world (and the astral-world) because the spirit-world cannot tolerate any form of impurity. The most efficient way to create impurity within the physical-world was to squeeze a small part of YOURSELF into a physical body and call it human. That is you. You are an impurity creator - a human *being* impure.

The human mechanism of impurity is negative thought. Thought is the source of everything that a human does. Thoughts drive your instincts, choices, skills, personality and imagination. Those thoughts then drive your actions, speech and movement. Thoughts and actions combined then create your feelings. Thought-action-feeling is then the sum total of all human behaviour - and it's all driven by thought.

To create impurity of thought as a human, YOU agreed that you would unwittingly develop a bad, subconscious habit of *reacting* with negative thought.

There are two ways to behave as a human - respond or react. You *respond* to life by consciously deciding what to think-do-feel. This is the basis of your wants, needs, choices, communication and creative-intellect. You *react* to life by subconsciously reacting based on the learned or memorised automatic instructions stored in the subconscious memory (the memory of how you acted before). This is the basis of your instinct, learned behaviour, personality and creative-imagination. Both behaviours are valid and necessary.

YOU knew that *reactions* operate beneath your conscious awareness (sub-conscious) as automatic, hidden, habitual forms of behaviour. In other words, you react without thinking about it. YOU knew the subconscious mind would then be a *perfect place* to develop a *bad, subconscious habit of reacting with negative thought*, mainly because that habit would be powerful and sneaky enough to overpower your default powerful and wise potential. That would then guarantee a regular, diverse, unique and evolutionary supply of impurity.

YOU set up your life so that you would learn most of your bad habits of negative thought as a child. YOU knew that when you were little (0-7 years of age), you would worship your parents as Gods. YOU knew you would never question their actions - why would you, you're only little. As a child, you just assumed everything 'the Gods' did was perfect.

For example, as a child, if mum or dad got frustrated and yelled at you, YOU knew that you wouldn't have the brain power to analyse the event and come to a sensible conclusion that mum and dad were just tired and stressed and didn't mean the comments, 'Hey, get down from there you idiot. You'll fall and hurt yourself!', or 'Not now, I'm busy', or 'Don't be silly/naughty/loud'.

YOU knew that your incorrect perception of parent perfection, combined with a limited child-potential and an overactive imagination, would often misinterpret such events as being *all about you* by thinking, 'The Gods (parents) are yelling at me (or ignoring me, belittling me, criticising me). They must be angry. But they're perfect and I worship them. So, it can't be about *them*, so it must be *about me*. There must be something *wrong* with me. I must be *lacking* in some way. I must be a *bad* person. I must be stupid, naughty, useless, clumsy ...'

YOU knew that you would also learn bad habits of negative thought from your parents, siblings and society. You would assimilate bad habits of negative thought from the negative culture of your society. You would inherit bad habits of negative thought from the belief systems of seven generations past (the sins of the father). These bad habits of negative thought would also make you addicted to negativity and develop a bad habit of focussing on negativity, making the bad habits of negative thought even worse.

These bad habits of negative thought would all accumulate in the subconscious mind. The rudiments of these bad habits would be developed from 1-7 years of age, making its fundamental structure childish and unrealistic. From 8-14 years you would strengthen these bad habits during puberty, making it immature and pubescent. From 15-21 years you would reinforce this bad habit during teenagehood, making it rebellious and irresponsible. By the age of 21 years, this bad, childish, unrealistic, pubescent, immature, rebellious and irresponsible habit of thought would become a permanent part of your subconscious personality.

Being subconscious, that bad habit would lie dormant until activated. Being negative habits of thought, they would be activated by negativity. If you *didn't* make a conscious choice to respond with power and wisdom to *overcome* those feel-bad emotions, you would be ignoring your powerful and wise potential. As soon as you ignore your powerful and wise potential, it creates an opportunity for those bad habits of thought to take over and urge your brain to think with negativity.

When they take over, those bad habits of negative thought make you act like a poorly-behaved child. It makes you selfish, greedy, angry, jealous, depressed, irresponsible, rude, etc. This list is very long (see table on page 168). It also feeds a bad habit of *focussing* on negativity and being *addicted* to negativity.

And being a subconscious habit, you would be *no more aware* of this habit than you are aware of any other subconscious habit. For example, when you drive a car, you are mostly completely unaware of the thousands of *subconscious tasks* required to drive that car. You just drive, unaware that your subconscious mind is constantly retrieving a whole bunch of old habits-of-thought to feed the actions required to look, listen, turn, brake, clutch, change gears, etc. That's because you have driven that car *so many times* that the thoughts required to do so have become repetitive, then habitual, then *automatic subconscious habits of reaction*.

In short, whenever you feel bad and don't choose powerful and wise thought, there is a very good chance you will *react badly* like a poorly behaved child, and you do so without consciously thinking about it or being aware of it. It will make you react badly to yourself, to other people and to life in general. It's the primary cause of all misery and falseness and is otherwise known as the ego.

Nobody deliberately chooses to think with insecurity, failure, depression and ignorance and then *feel* bad. It's not like you wake up one morning and decide, 'OK, today I'm going to struggle. I'm going to avoid the opportunities in life because of my insecurities. I will fail, achieve nothing and feel unworthy. I won't speak up, I will let people walk all over my boundaries and I'll feel miserable. And just to top off the day, I'll ignorantly judge people's actions and disconnect from my Soul. I'm going to feel like shit, all day. Yah!'

The ego is an essential part of your subconscious personality and human nature. It's a powerful force chosen by YOU, integrated *within* you and co-facilitated by everyone around you and its job is to urge you to create impurity as a human. It's how you define your level of GODlikeness (goldlikeness).

The ego has been called many things such as the alter ego, the false-self, the mask we all wear, the wounded inner-child, our coping mechanism, the devil within or essential-suffering. I prefer to call it the 'EGO' (bigger version of ego) because its purpose is way bigger than the psychological version of the ego.

The EGO is a huge topic and is covered in detail in Yoooge Book Two but basically the EGO breaks down into four bad reactions - fear, failure, depression and ignorance. These four reactions are the *exact opposite* of the four *responses* that make up power and wisdom - security, capability, expression and knowing:

CONSCIOUSLY *RESPONDING* WELL

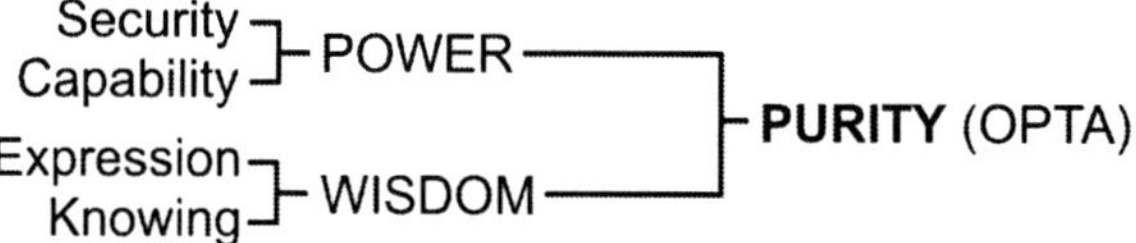

SUBCONSCIOUSLY *REACTING* BADLY

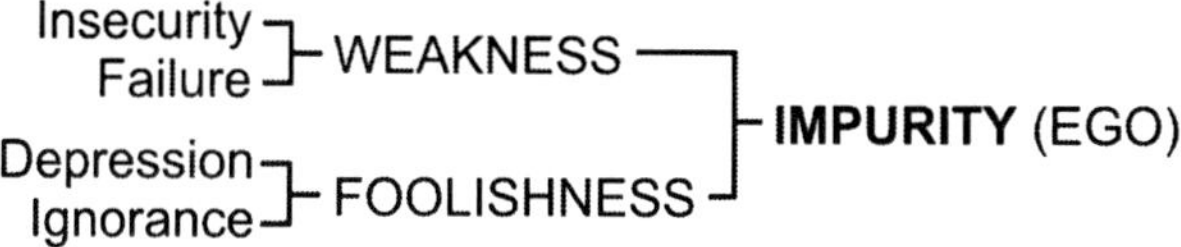

(Figure) The extreme-duality of purity and impurity demonstrating how the EGO is the exact opposite of power and wisdom (OPTA - explained on page 170).

Bad Reactions of **FEAR**	
You	Panic attacks, anxiety, fears, phobias, apprehension
People	Non-social, trouble meeting people, non-trusting
World	Lack of adventure, not willing to take risks, workaholic
Things	Over-competitive, greedy, jealous, hoarding

Bad Reactions of **FAILURE**	
You	Frustration, anger, self-criticism, hopelessness
People	Unreliable, insecure, annoying, not trustworthy
World	Lack of energy, willpower and strength; rely on others
Things	Mistakes, accidents, clumsy, unskilled, lacks talent

Bad Reactions of **DEPRESSION**	
You	Depression, boredom, apathy, anger, misery, feeling 'blue'
People	Can't say 'No', or stand up for yourself, or express emotion
World	Unimportance, can't change things, victim mentality
Things	Not artistic, creative or individual, don't like people

Bad Reactions of **IGNORANCE**	
You	Indecisiveness, confusion, regret, guilt, worry, stress
People	Criticism, judgement, stupidity, carelessness, selfishness
World	Soulless jobs, don't know your purpose, unfulfilled, stale
Things	Lack of creative imagination, don't know what you like

(Table) The bad habits of negative thought of the EGO that make you act like a poorly-behaved child

YOU AGREED IT WOULD BE VOLUNTARY

Does the EGO seem entirely unfair? No, because even though the EGO was a subconscious force powerful enough to guarantee a regular supply of impurity, YOU agreed that you would have *complete free will* to think any way you chose.

As a human, you would have to make a choice, *every time* you thought. You could at any point, *embrace* your powerful and wise potential and choose to respond to life with powerful and wise thought. This would make you think-act-feel with security, capability, expression and knowing. Alternatively, you could at any point, ignore your default potential and allow the EGO to take over so that you react to life with weak and foolish thought. This would make you think-act-feel with insecurity, failure, depression and ignorance. One would feel good, the other would feel bad. Simple!

In that regard, every negative thought would be a *voluntary* journey into impurity (impurity of thought) and every positive thought would be a *voluntary* journey into purity (purity of thought). YOU knew that you could change the direction of that journey, at *anytime*, between purity or impurity, just by changing your thoughts. YOU knew you always had that choice and the Cosmos would always respect that choice. This made the illusion fair for all humans, especially considering that any person can easily change what they think about, **now**. It's effortless and takes no training.

'Bullshit!' Or at least that's what a couple of friends yelled when they read that statement! Let me explain. Changing your thoughts *now* is easy. Try it. Think of the colour green, in your mind. Now think of the colour red. Now turn that red colour into an apple. Now place that apple in the mouth of a camel. Easy.

But your thoughts *now* are not the problem (mostly). They're the solution, sure, but not the problem. It's your old patterns of negative thoughts *from the past* that are the problem - your EGO. It's the EGO that makes you feel bad. It's the EGO that sabotages your attempts to be a better person. It's the EGO that keeps you trapped within a cycle of negativity so that you keep making the same mistakes over and over again.

The best way to fix a bad, subconscious habit of reacting with negative thought is by doing the *exact opposite* - a good conscious habit of responding with positive thought, which I refer to as Only Positive Thought Action – or OPTA for short.

The practise of OPTA is simple. No matter what happens in life, you make the conscious decision to *choose* positive thought *every time* - security, capability, expression and knowing (see 'The Four Thought Sets' on page 241). For example, instead of reacting with fear and lack, you choose the thoughts 'I am safe' and 'There is enough'. Instead of predicting a failure you choose the thoughts, 'I am skilled and have all the motivation I need to succeed'. Instead of doubting yourself, choose the thoughts 'I know what's best for me'.

Is that easy to do? No, it's a massive challenge! It's what you've come here to learn to do. It will take lifetimes to master. It's how you evolve. It's how you become more and more GODlike - a human *being* GODlike. It's what GOD is observing you do, over many lifetimes, so She can experience Her own Divine Magnificence by watching *you* rediscover your own Divine Magnificence.

You practise OPTA as much as you can before life eventually gets the better of you and you give up. Then you get to react badly, feel bad, go downhill, eventually cheer up, return to purity and start the whole cycle all over again - except each time you come back, you come back more powerful and wiser. That's the human journey - up and down on the emotional rollercoaster (discussed later).

So why bother? You choose OPTA not because positive thoughts will make you feel better - they do. Not because it's a better way of being - it is. Not because it will co-create a better life - it will. You choose OPTA because those newer more positive thoughts will accumulate in the subconscious mind and slowly *erode your EGO*.

Not only does that create a better life, make you a better person, create better relationships, heal disease, co-create more yoooge experiences and reconnect with your Soul; it reduces your bad, subconscious reactions, which reduces your feel-bad emotions, which *makes you feel better*. There is nothing more important than how you feel. Feeling good is *everything*, the highest version of which is joy (and love).

You may be here as a human to co-create impurity but part of that journey is to learn how to *maintain a constant state of joy*. That's actually one of the main reasons you create impurity. Impurity makes you feel bad. Humans don't like to feel bad. When you feel bad, you *immediately* seek to feel good, which is to seek joy. The thought-actions required to create joy are the *exact opposite* of the thought-actions required to react badly with EGO. In short, feeling bad actually 'forces' you to unwittingly resolve the EGO, whether you like it or not! OPTA is optional, evolution is not!

That's why YOU agreed to it all - the good and the bad stuff in life that you would have to experience, directly or indirectly. YOU knew the good stuff would make you feel good. You knew the bad stuff would make you feel bad.

YOU knew that the most intense experience for you as a human, would be *feeling*. It would either *feel* amazing or it would *feel* awful. YOU knew that would be a powerful motivating force. It would either feel amazing enough to encourage you to seek higher versions of joy, which would lead you to love, or it would feel *awful enough* to 'force' you to think-do whatever you could to feel good again, which would lead back to joy. Either way, the outcome is joy and the thought-processes behind joy would resolve the EGO.

YOU also knew the greatest outcome of life, your reward, your mark of success, would be joy and love.

YOU AGREED ON THE START POINTS

YOU knew that the *kind* of journey you would take would depend on your original start-point, determined by what you think about most. Again, that would *always be voluntary*. You would then journey downhill *from there* into impurity and eventually return back to purity.

Negative people always start a journey from a negative point. They can go downhill into impurity (more negativity) from there and eventually cheer up to return to a state of purity. Positive people always start their journey from a more positive point. They can go downhill into impurity from there and eventually cheer up to return to a state of purity. Again, that would *always be voluntary*.

Journeys *uphill* into purity are much less common. Humans have such an EGO and such a poor habit of consciously responding to life with positive thought (OPTA) that it's often difficult to be more positive than they already are. Most people normally maintain a default level of purity (purity of thought) and journey *downhill from there* into impurity (impurity of thought) and then back again to purity. If you're willing to change your thoughts, you can *start* your journey *whenever* you like and take *whatever* journey you like.

YOU AGREED TO FORGET THE TRUTH

YOU knew that in order to create an EGO, you would have to remove much of your GODlike powerful and wise potential. You can't entertain impurity of thought if your mindset is filled with GODlike purity. YOU knew that to remove your GODlike potential you would have to *forget you were a God,* since the only way to remove potential from the mind, is to forget. When YOU squeezed a small part of YOURSELF into the physical-world as a human-being, you also squeezed a *very small portion* of YOUR powerful and wise potential into your human mind and then made sure you forgot the rest. Part of the forgetfulness includes forgetting you are *one with all things,* and forgetting that the Cosmos is a giant illusion.

Consequently, as a human you've forgotten you're a God, so you believe you are a human-being. You've forgotten you are *one* with all things, so you believe you are *separate* from all things. You've forgotten the Cosmos is an illusion, so you believe *reality is real*. The combination of the three creates the unequivocal belief that you are a separate, human-being, living within reality. The Truth is that you are a God with amnesia, connected to all living things, participating in a giant illusion. This is what I summarise as **'forgetting the Truth'**(with a capital 'T' since it's a Divine Truth).

Forgetting the Truth is similar to forgetting the formulas of complicated mathematics a decade or two after high school. For most people, the memories are still there, locked away in the subconscious mind, it's just they're consciously forgotten. You can't remember, unless you *try* to re-member ('re-member' as in putting things back together, the opposite of 'dismember'). If you re-experience mathematics, it will probe your subconscious mind and the memory will return. You may have to re-learn a few things but most of the basic memory is still there and you will re-member most of your early mathematics.

Re-membering potential is very similar. As you experience the challenges of life, you have to *find the potential within you*rself to overcome those challenges. This searching for potential is what helps you re-member your forgotten potential. The purpose of extreme-duality is to help you re-member that forgotten potential because the greater the duality, the greater the contrast, the greater the experience, the greater the challenges, the more potential you will have to find, the greater the re-membering.

The purpose of life is to re-member your forgotten potential so you can attain higher and higher levels of potential. You then use those higher levels of potential to urge your brain to think with more security, capability, expression and knowing. This will make you *feel* more courageous, successful, fulfilled and purposeful. These feelings will combine together to create the emotions of desire and inspiration, which then combine together again to create the super-emotion of joy. The desire will drive you, the inspiration will guide you and the joy will make it fun. You'll naturally end up doing (desire) what you enjoy because you know (inspiration) what you enjoy doing. You'll naturally end up being the best you can be at what you enjoy doing the most. Sharing that joy with others, by serving others with your power and wisdom, is what turns that joy into love. Love is the final and most desired outcome of life.

The more lifetimes you live, the more potential you will re-member, the more powerful and wise your thoughts will become, the more joy and love you will feel, the more GODlike you will become. As you re-member more and more of your potential, you will also re-member more and more of the Truth that you forgot. Eventually, you will re-member, fully and completely, that you are a God, connected to all living things, participating in a giant illusion. This will take many lifetimes.

YOU AGREED TO THE GIANT MAGIC TRICK

YOU knew that you and most of the world would have *absolutely no idea* that they were Gods, connected to all things, participating within a giant illusion. Forgetting the Truth, combined with the added 'realness' of co-created vibrational-reality, would mean that very few people would ever question that reality.

Reality looks real, sounds real, tastes real, smells real and feels real! There are historical records and fossils that show we have a real past. Everybody else *thinks* it's real. Scientists can even interfere with the outcome of their own observation and *'prove'* it's real. Why would you question that? Who questions that? Part of our genetic program of the illusion is *not* to question that.

If you did, especially a few decades ago, well-meaning people would sneak around behind your back, ring 1800-My-Friend's-Gone-Bonkers and organise an unmarked van, filled with well-meaning psychologists to come around to your house, pick you up, pump you full of powerful 'nighty-night' pharmaceuticals and tuck you away in a nice, warm, padded room!

'Hello, 1800-My-Friend's-Gone-Bonkers. How can we help you?'

'Hi! We need help!'

'What seems to be the problem?'

'We have a family member who's completely lost his mind'.

'Okay. Are you sure?'

'Yes, completely. He keeps going on and on about how the world *isn't real!'*

'Really?'

'Yes, I knooow. Craaazy right?'

'Okay, we'll send a van-load of well-meaning psychologists around to your house, right away!'

'Oh thanks'

'No problem. You snub em, we drug em!'

'Awesome!'

Nevertheless, reality is an illusion; a gigantic magic trick; a really good one, engineered by GOD and facilitated by Gods and as with any good illusion, we've all been fooled. For example, in a magic show, we know a magician can't saw his beautiful assistant in half with a handsaw, as she lies smiling inside a tinsel covered box with her head poking out one end and her feet poking out the other. There would be blood – pools of it. As impressive as the illusion is, we know when she climbs out and takes a bow it's a trick and it's not real.

We know it's an illusion. We pay money to go see the illusion. So why then are we on the edge of our seats as the illusion is presented to us? Why does the illusion shock and tantalise us so much? Why did we feel so *alive*?

Why? Because there was a moment in the show when we *believed* it was real! Especially with the powerful atmosphere of stage lighting, music, the energy of the crowd and the fake screams of the assistant. We were fooled.

Well guess what? We live *within* one of the biggest and most complex magic tricks of all time (and space) and *we're still inside the theatre* watching the illusion because we forgot the Truth. We're now so completely transfixed by the light show, so 'on the edge of our seats' with reality, so caught up in our daily lives that we never question whether that reality is real or not.

When was the last time you and your friends, sat down for dinner and then asked the question 'Is this dinner plate real? Is the chicken just a trick of light? Are those potatoes just vibrational-reality, held together in form, function and phenomena by the electromagnetic forces of the Cosmic-pond? Is the … (nom, nom, nom)?' Nom, nom, nom is the sound of your disinterested friends gorging themselves on chicken as they ignore with disbelief your tyrannical rave about metaphysics. It's programmed into their DNA.

What's really interesting is if you combine the words 'spiritual' and 'reality' and then remove the word 'real' from the middle, it forms the word 'spirituality' (spiritual_real_ity). This is especially interesting when you consider that life is a spiritual journey within a reality that's not real. Nom, nom, nom …

The Australian Aboriginals were right all along (of course!) *The Dreamtime* is our world, which is the illusion, whereas our dreams (being part of the spirit-world) are the reality. Tell that to the well-meaning psychologists!

But what your chicken-gorging friends may not realise is that once you can accept that life is an illusion, you can play God! Knowing that the world is an illusion and knowing how the illusion operates, enables you to co-create a *you*niverse that matches your desires and inspiration. You play God because YOU are a God and YOU do the actual creating. You just choose what you want via your thoughts. Together, YOU and you can slowly, eventually and gracefully co-create the life of *your* dreams. This is the basis of Yoooge Book Three.

Knowing the world is an illusion also helps you understand human nature enough to forgive, be grateful and express compassion. It stops you worrying about death. It helps you connect with your Soul. All of that creates more joy, joy feels amazing and when you feel good, it makes you want to *be here more.* That, strangely enough, grounds you *even more* within the prospect of reality.

Although, make sure you tell your chicken-gorging friends we all have to *live within* the constraints of the illusion, whether we know it's real or not. The illusion still operates on a very 'real' electromagnetic platform that will slap your face into the concrete every time you jump off a building. Even if you watched 'The Secret' and thought you could fly. Vibrational-reality can still take the form, function and phenomena of a bus and squish you flat if you cross the road, drunk on a Friday night. Depression still kills teenagers and love still makes you worry about your kids! You can co-create like a God but you can't circumvent Cosmic Law (not yet anyway!).

YOU AGREED TO LIMIT YOUR SENSES

Like any good magic trick, YOU agreed to set up a 'screen' to hide the illusion of the magic trick from you, so you couldn't see how the magic trick works. YOU did this by limiting your human senses to within the boundaries of the illusion. This meant as a human, you could only perceive the illusion itself, but not beyond it (I am generalising here, as many psychically sensitive people can perceive beyond the boundaries of the illusion, including the following 'limitations').

YOU limited your human hearing to 20-20,000 hertz, stopping you from hearing the vibration of Mother Earth (a deep rumbling 8-9 hertz frequency), spiritual communication (clairaudience), the spheres or the Angelic symphonies (described by some as the most beautiful music imaginable) or the atomic buzz of electromagnetic vibration.

YOU limited your human sense of smell to physical-world smells and tastes, stopping you from smelling the astral-world (people have reported familiar smells from deceased relatives such as perfume, tobacco or incense), or the intoxicating perfume of electromagnetic vibration.

YOU limited your human sensation of touch to feel solid, liquid and gas, stopping you from feeling the subtle energy fields of the human aura, mind-energy of the astral-world or the Gods within the spirit-world (the Gods are not above you in the 'Heavens', they're all around you, even *within* you).

But most importantly, YOU limited your human eyesight to the spectrum of *visible* light. As a human, ninety to ninety-five percent of how you perceive the illusion is via your eyes. If you turn the lights out, it's dark and there is 'nothing there', so to speak.

Reflections, refractions and absorption of light are what give the physical-world shape, shadow, colour, texture and luminescence. The light we see seems like it has three dimensions because the light waveform is separated by space-time. The light seems real because the electromagnetic forces within the Cosmic-pond holds it all together in form, function and phenomena. The light itself fools us because it's intelligent and alive and morphs to suit the collective and individual requirements of the Cosmos.

But it's still only light and it's still only what we can see. There's a lot more going on behind the scenes that we cannot see. Ninety-nine percent of the light spectrum is hidden from us as ultraviolet light, infrared, radio waves, microwaves, X-rays, gamma rays and cosmic radiation. In fact, if visible light was an octave on the keyboard of a piano, the full light spectrum would represent a keyboard ninety-three million miles long. That would stretch all the way from the Earth to the Sun.

Even the full spectrum of light is only a very small part of the full spectrum of electromagnetic vibration that stretches way past visible light and into God-Light (astral-world, the Cosmic-pond and the spirit-world). We can't see any of that either. If we could, we would be astounded at what's going on behind the scenes.

Most noticeably would be that the light is flickering on-off as part of the pulse created by YOU and all the other Gods. We humans can't see that pulse because it buzzes at such a high frequency. You get the same effect with fluorescent lights. Fluorescent light bulbs flicker on and off at approximately one hundred times a second but that's too fast for us to notice. The millisecond after-glow of the light registering in our eyes is longer than the on-off cycle of the fluorescent light bulb and so it always appears as if the light is 'on'. Similarly, we have absolutely no chance of spotting the nonillion pulse of the Cosmic-pond and so reality always appears as if the light is 'on'.

YOU AGREED TO THE MINOR JOURNEYS

There are two types of journeys into impurity of thought. There are minor journeys into impurity and back again, and there is one major journey into impurity and back again. As a human, your minor journeys into impurity are done via your daily thoughts. Every negative thought is a voluntary journey into impurity. From there, every positive thought required to cheer up is a voluntary journey *back* to purity.

YOU agreed that *you* would facilitate regular, minor journeys into impurity via your human thoughts. **Thoughts create your feelings.** If you ignore your potential the EGO often takes over. The EGO makes you feel bad and so you go downhill. You don't like feeling bad. No human does. You eventually cheer up. All humans do. That's also programmed into your DNA. Humans will do *anything* to feel good and humans will *avoid everything* that feels bad.

The cheering up process takes as long as it takes you to embrace enough powerful and wise potential to think better thoughts to make you feel better again - or at least feel the same way you *did* before you went downhill. That can take a few minutes, hours, days, weeks, months or even years. We've all had a bad day, or week or many of us have had a bad month or even a bad year.

How fast you cheer up depends on how fast you can change your thoughts. You're the only one in charge of your mind, which means you're the only one in charge of your thoughts, which means you're the only one in charge of *how you feel.*

To cheer up, you have to find the potential within your mind to think thoughts that are *more positive* than the ones that make you feel bad, in order to feel better again. This searching for potential 'forces' you to either re-member forgotten potential, strengthen existing potential or find newer, higher levels of potential.

Consequently, whenever you cheer up, you usually come back from your voluntary journey into impurity with a slightly higher level of powerful and wise potential, to whatever degree matches the depth of your impurity. Little trips into impurity create small increases in potential. Big trips into impurity create great increases in potential. That's why YOU agreed that you would live in a world of extreme-duality.

When you complete your journey and return to purity, your new levels of potential get stored in your subconscious memory and become a *permanent* part of your subconscious personality. That means you take a *permanent* step towards purity. Potential translates to Light. These slightly higher levels of potential in your mind create more Light and so you also come back from each minor journey into impurity, slightly en-**Light**-ened.

That's why YOU agreed that you would live in a world of extreme-duality. The greater the poles of extreme-duality, the greater the contrast, the greater the experience, the greater the challenge - the more powerful and wise potential you need to find *within yourself* to overcome that challenge, the more potential you re-member, the more en-Light-ened you become. In short, extreme-duality en-Light-ens you.

YOU AGREED TO THE MAJOR JOURNEY

YOU knew that your minor journeys into impurity would help you re-member more and more potential, slowly. Over the period of your lifetime, this slow re-membering of potential would evolve your mindset so that you become more en-Light-ened. YOU agreed that as a human, this evolving mindset would form the basis of one major life-time journey into impurity and back again.

That journey would start at birth and continue until you die. It would be done by most people, unwittingly and as part of a 'normal' life - unwitting since they forgot the Truth and so life is mostly lived unwittingly. Some would pursue personal growth and spiritualty for faster en-Light-enment but most would evolve naturally as part of the default setup of the illusion. The journey would unfold in cycles of seven years and twenty one years (3 x 7 year cycles) and the normal duration of the major journey is one hundred and five years in length.

The first part of your major lifetime journey would be a twenty-one-year long *journey into impurity*, by developing the EGO. The fundamentals of that habit would be learned in childhood from 1-7 years. You would reinforce and confirm your EGO from 8-14 years during puberty and again through 15-21 years of teenagehood (yes, it's a word used more and more in urban language but not a valid Scrabble™ word!).

YOU knew it would take twenty-one years to fully develop your EGO. It would then be a fundamental part of your subconscious personality. It would lie dormant in the background, ready and willing to react, twenty-four hours a day, seven days a week. It would react when activated by negativity, but only after the choice was made *not* to respond with power and wisdom. You would then react badly, like a poorly behaved child.

You would probably *have no idea* you were reacting badly. It would also feed a bad, subconscious habit of *focussing* on negativity and being *addicted* to negativity. It would also ensure *regular* trips into impurity and create all the extreme-duality you required. It would also unwittingly en-Light-en you as it forced you to seek joy, which is to seek OPTA (only positive thought action)

The second part of your major lifetime journey would be an eighty-four year long *journey back to purity*, as you slowly resolve the EGO from the ages of 21 to 105 years.

From 22-28 you would unwittingly live the consequences of your EGO, thinking negativity and really *feeling* the reactive consequences of impurity of thought. In other words, before you would actively seek OPTA, you would first need to *immerse* yourself in EGO, otherwise why would you seek OPTA? This may not be a positive stage of your en-Light-enment but it is a necessary stage. The ages of 22-28 would also tend to be the most intense years of your life - good and bad - especially 28 years.

From 29-35 years you *begin* to resolve the EGO. This marks the start of your true en-Light-enment. It's like you wake up on your 29th birthday and decide, 'From today, I am going to think and act with more power and wisdom so I can feel more joy'. Not that literally of course. It's more like you have this mini mid-life-crisis where this inner urge makes you try to be a better person *in some way*.

Some people feel a strong urge to question their relationships or their own happiness. Some people feel a powerful desire to conquer their fears, finish a task or go back to study. Some people finally become inspired enough to follow their passion as a new career. Some people suddenly feel like it's time to have a baby, whilst others just know it's time to change so they can be a better parent. Some withdraw from life with drugs and alcohol or sink into deep depression. Some people seek a guru, others do a course, many read a spiritual-type book and some even travel to India to find god.

Some experience this at 28 years, most at 31 years, some never change but usually by the age of 35, a new cycle of en-Light-enment is well under way. That en-Light-enment becomes more intense from 36-42 years, is integrated from 42-63 years, and becomes part of you by 63-84 years. As you can see, the Cosmos is not in a hurry!

Your en-Light-enment culminates when you become a wise elder at 78-84 years of age. From 63-84 years you pass on your power and wisdom to your children and grandchildren as the 'loving grandparents', to help them find *their* way back to purity and ensure each subsequent generation evolves. At 84 years of age YOU choose to 'go again' and begin a whole new cycle of purity from the ages of 84-105 years, or YOU choose a transition back to the spirit-world via death.

YOU knew that en-Light-enment unfolded in cycles of seven and twenty-one years (3 x 7 year cycles) and so YOU agreed to a lifespan of five rounds of twenty-one years (5 x 21 = 105 years). This would be enough time to completely resolve the EGO but *only* if your mindset entertained enough positivity to keep you healthy enough so that you actually reach 105 years of age (life-force flows from positive thought).

YEARS	NICKNAME	THY SELF	STAGE OF DEVELOPMENT
1-7	Make Up	**False Self**	Create False Self
8-14	Blow Up - Expand	**False Self**	False Self Midpoint
15-21	Grow Up - Man	**False Self**	Know False Self
22-28	Play Up	**The Self**	Define False Self
29-35	Shake Up	**The Self**	Self Midpoint
36-42	Wake Up	**The Self**	Know Self
43-49	Break Up	**True Self**	Define Self
49-56	**Blow Up (Mid Life)**	**True Self**	**True Self Mid Point**
57-63	Open Up	**True Self**	Know True Self
64-70	Give Up (Wants)	**Higher-Self**	Define True Self
71-77	Love Up (Grandkids)	**Higher-Self**	Higher-Self Midpoint
78-84	Wise Up (Elder)	**Higher-Self**	Know Higher-Self
85-91	Start Up (New Cycle)	**Jesus-like Self**	Define Jesus-like Self
92-98	En-joy	**Jesus-like Self**	Jesus-like Self Midpoint
99-105	Light Up	**Jesus-like Self**	Know Jesus-like Self

(Figure) Table showing the Cycles of Seven as we come to 'know thyself'. The ages of 105-126 years of age are a bit of a mystery because not enough people have lived that long!

YOU AGREED TO DEATH

YOU knew the journey would end abruptly, guaranteed by death. You could then return home to the spirit-world and re-join you with YOU. YOU knew there would then be 'one hell of a party' when you returned home, as all the other Gods celebrate your contribution as an impurity creator. The special guests of that party would include all the Cosmos, the First Being, the Gods and GOD.

GOD would then 'hug' you with absolutely mind-blowing love and gratitude for helping Her experience Her own Divine Magnificence. She would 'tell' you how She used her infinite viewpoint to observe all the powerful and wise potential behind your thoughts, from all your lifetimes, so that She could *be* love. She would tell you how She collectively observed every other human doing the same and used that collective viewpoint to have a disconnected, impartial experience of Her own Divine Magnificence.

As a reward for helping GOD, She would invite YOU to experience the feeling of being GOD. YOU would then experience all love from all beings from all time!

That's the agreement YOU made. That's why part of YOU squeezed itself into the physical-world as a human *being* impure. That's why you're here, *now* on Earth, reading this book.

SUMMARY

- YOU are a God.
- YOU made an extensive agreement to help God experience Her own Divine Magnificence.
- YOU knew the Cosmos was too perfect.

- YOU agreed to add impurity to the mix.
- YOU knew this impurity would help present an expanded and complete view of GOD's Divine Magnificence.
- YOU agreed to incarnate as a human and co-create impurity of thought – negativity.
- To create impurity as a human, you agreed to develop an EGO.
- The EGO would operate as a bad, subconscious habit of reacting with negative thought and guarantee a regular, unique and evolving supply of impurity.
- To make things fair you would have free will to think anyway you like.
- You could *embrace* your powerful and wise potential and choose to respond to life with powerful and wise thought. This would make you feel good.
- You could ignore your default potential and allow the EGO to take over so that you react to life with weak and foolish thought. This would make you feel bad.
- Every negative thought would be a *voluntary* journey into impurity (impurity of thought) and every positive thought would be a *voluntary* journey into purity (purity of thought).
- If you're willing to change your thoughts, you can *start* your journey *whenever* you like and take *whatever* journey you like.
- YOU agreed that you would forget the Truth.
- YOU limited your senses to the boundaries of the illusion.
- You rarely question the illusion of reality - who does?

- YOU agreed that you would take regular minor journeys into impurity and back again - hourly, daily, weekly, monthly and yearly.
- Each time you ventured into impurity you would have to find the potential within your mind, to think thoughts that are *more positive* than the ones that make you feel bad, in order to feel better again. This searching for potential 'forces' you to either re-member forgotten potential, strengthen existing potentials or find newer, higher levels of potential.
- YOU agreed to one major journey into impurity of 0-21 years and then back to purity for the next 22-84 years, if you live that long. The cycle would start again from 85 years.
- YOU agreed you would read this book.
- YOU agreed that you would read this summary - fully (good for you!).
- YOU agreed the journey would end abruptly with death.
- There would be a big party in the spirit-world for you after you die (woot, woot).

YOU SET UP YOUR LIFE

Anything challenging in life (good or bad) will 'force' you to find the potential within yourself to facilitate, master or conquer that challenge. This either strengthens existing potentials or forces you to re-member new potentials. Consequently, the Gods set up life to be quite challenging - for everyone.

The natural challenge of life then ensures a default level of en-Light-enment for all people, making the illusion fair for all people and guaranteeing a journey back to purity for all people. Everyone evolves, everyone resolves the EGO and everyone becomes en-Light-ened - whether they like it or not - and for good reason. It would be unfair if some knew the way of en-Light-enment and others did not, especially considering the cycle of en-Light-enment can be eighty-four years long. Therefore, a certain level of en-Light-enment is automatic.

THE SOUL-GROUP

To help set up the challenges of life, your Soul got together with a group of other Gods in the spirit-world to form a Soul-Group. The role of the Soul-Group is to plan and choose all the major challenges in your life. Consequently, your Soul-Group chooses your country of origin, culture, upbringing, destiny, legacy and finally your own death (ironically, the biggest challenge of *life*).

But nothing challenges you more than human relationships - good or bad. Personal challenges require the greatest amount of potential and hence offer the greatest level of en-Light-enment. Your Soul-Group plans and helps facilitate all the human interaction in your life. Your Soul-Group chooses your parents, siblings and extended family. Your Soul-Group chooses your friends, social set, colleagues and mentors. Your Soul-Group also chooses all the human interaction that leads to events that seem to be *outside of your control* such as drama, problems and accidents.

To facilitate all these wonderful personal challenges, various members of your Soul-Group then incarnate together as *humans,* with *you*. All the people in your life are incarnated-human-members of your Soul-Group. The people you're *closest to* such as your family, friends and partner usually represent the inner incarnated-human-members of your Soul-Group. The people in your life that have had a direct impact on you, such as your mentors, colleagues and your foes, usually represent the outer incarnated-human-members of your Soul-Group. The people who have served you, assisted you and even affected you as 'strangers' usually represent the very outer incarnated-human-members of your Soul-Group.

Each member of your Soul-Group then challenges you on a very specific and unique level. Your family challenge you on a very personal level. Friends challenge you on a personal-social level. Mentors, foes and colleagues challenge you on a professional-personal level. Strangers challenge you on a collective level. The challenges of human relationships in conjunction with the challenge of life itself then ensures a default level of en-Light-enment, for you *and* everyone in your life.

PURPOSEFUL CHALLENGES

But that's not enough! Default en-Light-enment may well en-Light-en you but it's *slow.* Your Soul expects *you* to *voluntarily* challenge yourself, as well. You are required to choose your *own* challenges and challenge yourself in a demanding, exciting and enjoyable way - a purposeful challenge.

A purposeful-challenge can be an enjoyable hobby, a demanding but rewarding career, exercising to lose weight, being a better person/parent/friend/mentor, raising a family, serving others, etc. There is no perfect purpose for everyone, just a perfect purpose *for you*. It doesn't matter whether you create a billion-dollar-a-year business that changes the world or knit belly-button fluff into a hammock, as long as it's challenging (demanding, fulfilling, rewarding, etc) and enjoyable (most of the time). That's what makes you feel alive! That's what causes life-force to surge through your body.

Purposeful challenges (although this may be stretching the term 'enjoyable' for some people) may also include forgiving family and friends, loving unconditionally, quieting the mind, or resolving the EGO.

A great historical example of a purposeful challenge is the first woman Pope. There is a highly contentious story about a girl who was so frustrated by the lack of educational opportunity for women that she disguised herself as a *boy* in order to enter a Benedictine monastery. She studied to become 'Brother' John Anglicus, then a curial secretary, then a cardinal and was eventually elected pontiff in 853, after the death of Pope Leo IV.

Everybody thought she was a man and the common dress of the Middle Ages allowed her to continue the guise. She reigned for nearly three years as Pope Joan (a common man's name of that age but oddly ironic) … as a 'man'. In that era, *that* type of challenge would have been super exciting, demanding and enjoyable (most of the time) but also extremely rewarding and life-changing. She changed history!

She also changed official Church procedure! A few hundred years later, a special chair was made with holes drilled underneath. New Popes were then required to sit on this chair naked, whilst a special committee of Cardinals peered through the hole from beneath to determine the Pope's gender.

Unfortunately, whilst mounting a horse during a Papal parade, she went into *labour* and gave birth. Angry parishioners tied her feet to her horse's tail and then dragged her to death.

The Church denies any such incident took place (of course!), although it's interesting to note that the story has been made into two films and numerous books and plays - most notably the Donna Woolfolk Cross's 1996 novel called 'Pope Joan'.

This doesn't mean your challenges have to be as extreme as Pope Joan's, especially considering her purposeful challenge killed her! The reason why purposeful challenges are so important is because they require a person to *consciously* choose powerful and wise thought, to facilitate, master or conquer the challenge.

Just look at Pope Joan's challenge. That was the greatest Papal practical joke of all time. *Imagine* the courage (security), skill (capability), verbal skill (expression) and knowledge (knowing) required to pull that off! What a challenge. What fun! What potential! That challenge would have 'forced' Pope Joan to embrace every possible kind of powerful and wise potential, strengthening existing potentials and re-membering higher and higher levels of *new* potential. Imagine the en-Light-enment she would have experienced, especially in that era. That woman/man was a genius.

Whether you're trying to make the perfect cup of tea or masquerading as a she-man-Pope-figure, a purposeful-challenge is the enjoyable path to en-Light-enment. It's how you're *supposed* to evolve. It's the innate urge within all of us to be a better person; to change, grow and expand; to learn, improve and get bigger; to renovate, upgrade and upsize.

People think you have to travel to India to find en-Light-enment. Others believe you have to chant, stand on your head and eat grass. Some seek en-Light-enment through god. Some seek en-Light-enment through a guru. But en-Light-enment comes from *you* seeking potential, *within* yourself. How many times have you heard the phrase, 'En-Light-enment comes from within'? It's not *within* India. It's within your own mind.

To seek out and find en-Light-enment, all you have to do is look within your own mind. There you would find the male-female God/Goddess of *power and wisdom*. That's the *only* deity you need to worship - your *own* potential. Contemplate your own security, capability, expression and knowing. Ask yourself, 'How powerful and wise *am I*?', 'Do I practise the positive thoughts on page 241?', 'What do *I* think-do-feel when faced with fear, failure, depression and ignorance', 'How much *joy* do I feel?'

If you can do all of that with profound gratitude and in a calm state of joy, then you have *already reached* en-Light-enment because *that's all there is* and it can only be done *now.* There is nothing to learn. There is nothing to practise. There is nothing you *have* to have, do, be, in order to achieve en-Light-enment. There is only what you think about *now,* how powerful and wise those thoughts are, and how much joy that makes you feel. That's all you have to do - embrace your potential, by *consciously* deciding that you will *respond* to life, every time, with positive thought (see page 241). You can do that at home, on the couch, in your own mind, whilst eating Indian take-away curry.

And as for those people travelling to India to find god, the Gods told us to tell you that god's not there. GOD is imperceptible nothingness, god does not exist, the Cosmos is a hidden framework of vibrating space-time, the First Being is beyond most humans and so the *only* deity worth worshipping is God/Goddess and that's within your own mind. So by all means, go to India for the cultural experience and a great curry, but not to find en-Light-enment.

THE COSMOS SUPPORTS PURPOSEFUL-CHALLENGES

The entire Cosmos supports a purposeful-challenge. It not only supports it, it helps you co-create it. As soon as you decide on a purposeful-challenge, the entire spirit-world springs into action. Your Soul instantly creates that event for you and sends it to you. That event will surround your youniverse. As soon as your thoughts create a vibrational environment in your youniverse that matches the vibrational-reality of the event, it will enter your youniverse. The magnetic-potential within your thoughts will attract that event to you and the electrical-potential within your thoughts will switch that event on.

At the same time, your Soul-Group will inspire the minds of *every human member* of your Soul-Group. If they can help you they will unwittingly be 'called' by their respective Souls to assist you. Suddenly, strange coincidences, synchronicities and opportunity will appear 'out of nowhere', into your youniverse, instigated by these people, from *your* Soul-Group. Family, friends, colleagues and strangers will 'miraculously' appear bringing with them all sorts of ideas, assistance, materials, opportunity and events to help you achieve your purposeful-challenge. It happens all day, every day, to everyone, everywhere and everywhen. You just have to look for it.

For example, if you decide to challenge yourself by losing weight, your Soul might inspire you towards finding a great fat-free menu book. You might feel a sudden inner strength or fitness-determination from your Soul. Your Soul-Group might organise a phone-call from an old friend who lost 45 kg on a great weight loss program. Your Soul-Group might send you a chance meeting with a fitness trainer who hands you a free one month membership to the local gym. Your Soul-Group might organise a series of 'coincidences' that leads to a new friend who also wants to lose weight (a fat buddy*).

*Only people who are overweight or have been overweight can use the word 'fat'. Skinny people have not earned the right to use that word. I have earned the right to use that word.

NECESSARY CHALLENGES

En-Light-enment is not an option. En-Light-enment is compulsory for all humans. We all re-member potential; we all become more powerful and wise. We all think with more power and wisdom. We all feel more joy. We all return to purity. YOU made that agreement with GOD, the Gods and your Soul-Group, before you were even born. How you do that, is *always* up to you.

The preferred method of en-Light-enment is via purposeful-challenges. This is the natural and enjoyable path to en-Light-enment. The choice to choose a purposeful-challenge is yours and the Cosmos will *always* respect that. It's your choice, it's your first option, and it's your birthright.

If you choose to ignore that choice and *not* purposefully challenge yourself, *often,* then your Soul will take over. To ensure an appropriate level of en-Light-enment, your Soul will challenge you, *for* you, in a necessary way - a necessary challenge. This is the 'forced' path of en-Light-enment and it often involves unnecessary pain and suffering (essential suffering).

Forgetting about wisdom for a moment and just focussing on power as an example, within the potential of power are the smaller potential-aspects of security and capability. Thoughts urged by the potential of security will give you the courage and the trust needed to 'give it a go'. Thoughts urged by the potential of capability will activate the skills and willpower required to achieve your goals. The combination of security and capability gives you all the 'tools' you need to succeed. If you choose to embrace these potentials, by *consciously deciding* to think in a courageous and skilful way, then you will most probably succeed.

If you often fail it's because you are, in some way, ignoring your security and/or capability potentials. You haven't made the choice to consciously choose powerful thought. This creates an opportunity for the EGO to take over and influence you with bad, *subconscious* habits of failure-based thought - especially when you feel bad.

You know when the EGO has taken over because there's this part of you (the EGO) that sabotages your efforts. It might be that you give up too soon, not complete the task, make foolish errors, feel a fear of failure or you might have a complete 'Oh shit, it caught on fire!' melt-down-disaster.

The EGO has taken over and subconsciously reacted with old, outdated habits of fear-based and failure-based thoughts such as 'I fear success', 'I don't have enough money/time/materials/help to succeed', 'I don't trust people', 'I'm not good enough', 'I can't do it' or 'I am not worthy of success'. The worst part is - you have no idea you're doing it!

Your Soul watches over you, always. It knows how powerful and wise you are, it knows how powerful and wise you can be and it knows how powerful and wise you *need to be* in this lifetime. It sees that you are stifling your own en-Light-enment by ignoring your power. It has sent you many desires and ideas that involve powerful challenges. It sees that you have not chosen to follow those urges and choose a purposeful-challenge. As a last resort, your Soul then sends you a failure-challenge to help you re-member your powerful potential, otherwise called a 'problem'.

Your Soul sends you this failure-challenge, knowing that you will fail. And you fail. In a bad mood, instead of responding to that problem by embracing your security and capability, you react to the problem like you always do, based on a bad, subconscious habit of reacting with negative thought you learnt as a child - failure. That failure-EGO overpowers the more subtle security and capability potentials, with fear-based and failure-based thought patterns. Consequently, you fail.

Your negative thoughts make you feel bad and you go downhill on a *voluntary* journey into impurity. You don't like to feel bad, no human does, so you cheer up, *eventually.* It could take minutes or months but you will eventually find the potential within you to think better thoughts and feel better - to whatever degree makes *you* happy.

If fear and failure made you go downhill then you will *feel* fear and failure. To cheer up from fear and failure you have to think thoughts of security and capability. You might think security-thoughts such as, 'Well, you know what, at least I had a go', 'I did the best job I could with the materials/support/tools I had', or 'It was great that others tried to help me'. You might think capability-thoughts such as 'I tried my best', 'I learnt from my mistakes' or 'I learnt new skills'.

To think those thoughts, you *have* to embrace existing security and capability potentials, or re-member new levels of security and capability potentials - potentials that YOU have purposefully ensured *that you* have forgotten. Your thoughts will make you feel better and the moment you feel better is the moment you have *overcome the challenge*. You then take a voluntary journey *back* to purity of thought.

It doesn't even matter if you don't solve the problem because the real challenge is *cheering up again* - more precisely, consciously choosing to find the potential within yourself, to urge thoughts powerful and wise enough to cheer up again. Cheering up from a bad negative reaction often requires *more* potential than solving even the biggest problem. It's almost as if the events of life are irrelevant. You do however get an extra potential boost from the success of solving the problem.

Consequently, you *come back* from your voluntary journey into impurity with a slightly higher level of security and capability potential. The potentials you strengthened or re-membered then match the potentials you are choosing to ignore. These newer potentials become a permanent part of your subconscious personality, making you slightly more powerful and ensuring you have more chance of success and less chance of failure, in the future. Da Daaar! The circle of life.

NECESSARY- CHALLENGES AREN'T BAD

Necessary-challenges can even happen *within* purposeful-challenges to ensure there is enough en-Light-enment to handle the newer purposeful-challenge. For example, if you decide to foster a child (an enormous yet extremely rewarding purposeful-challenge), yet still refuse to embrace the potentials and think the positive thoughts required to *manage* that challenge, your Soul might send you a necessary-challenge where you lose your job. This forces you to find a new job (which YOU knew that you would get) with a better income and flexible hours. The increased income helps you support your foster child. The flexible hours allow you to spend more time with your foster child. The demanding customer service of the new job teaches you to be more patient with your foster child.

People often judge necessary-challenges as 'bad'. But they're not bad. It's only really human judgement that labels a necessary-challenge as bad. Necessary-challenges are *necessary* because you don't challenge yourself enough, purposefully. If you're not challenging yourself purposefully then you're not fully embracing your powerful and wise potential. If you ignore your full potential then it creates an opportunity for your EGO to take over. When your EGO takes over, it can *really* take over.

The EGO is a subconsciousness force. It's the subconsciousness (non-awareness) behind the EGO that makes you react badly, yet you have no idea you're reacting badly. You're 'asleep' - reacting to life in a semi-conscious state. For the same reason, when your EGO takes over, you often have very little awareness that it has taken over. The EGO then feeds on itself. The negativity of the EGO makes you react badly. This makes you feel bad, which activates more EGO, which makes you react even worse, *more often* and so on … except you have no idea you're doing it.

Without conscious awareness you can easily and very quickly get stuck in a semi-conscious, negativity feedback-loop. You then keep making the same mistakes over and over again, all the while ignoring the potential required to remedy the negative thought patterns behind the mistakes.

To snap you out of it, your Soul sends you a necessary-challenge that's often drastic, 'in your face' or shocking enough to potentially *wake you up* from your semi-conscious EGOic state. That's why necessary-challenges often involve drama, problems, unwellness or stress. That's why people judge them as being bad. Drastic yes, but not bad and they're only drastic because you're not willing to challenge yourself, purposefully.

For example, if you don't embrace the potential for expression then you will not express yourself fully. If you don't express yourself, especially what's important to *you,* your ideas, your limits, what you expect from others and what sort of behaviour you won't tolerate from others, no one will know your boundaries. If people don't know your boundaries, because you haven't told them, they will 'walk all over you' by infringing on your personal space. When people infringe on your personal space, it will make you angry. If you don't express that anger it will lead to depression.

Until you find the potential within your mind to cheer up, the EGO will keep you locked in this negativity-feedback-loop. React, feel bad, react more, feel worse, etc. The EGO is feeding on itself to make your challenge greater. Metaphysically speaking, that's its job. It does that job very well and it does so subconsciously. This is the subconscious force that causes you to 'stew on your anger', which keeps you locked into a cycle of ever deepening depression.

The first stage in cheering up from depression is anger. Anger is a violent form of expression, or a whole lot of hurt expressed *all at once.* It takes a tremendous amount of potential (courage, trust, will, expressive urge and life-experience) to do that in a state of miserable depression - especially if you've been trapped in a cycle of depression for some time.

The reason why anger is a violent form of expression that takes a tremendous amount of potential to express, is because the potential and the expression required to get angry heals depression.

The EGO will make you feel worse and worse until you can't stand it anymore and you somehow find the potential within you *to express your anger*. Some people stand up for themselves. Some people snap and hurt themselves or others. Some people scream, yell or smash a dinner plate on the ground. Most people dislike the conflict of getting angry or standing up for themselves and instead unwittingly choose *other forms* of expression such as art, music, or exercise to 'express their anger' and counteract the feel-badness of their depression.

Either way, these people are expressing that anger and the potential for *expression* is strengthened and/or re-membered. This higher level of expression-potential will then *eventually* feed the urge to express their personal boundaries and avoid suppressed anger and depression. The circle of life. It's drastic but it works.

You can avoid all of this by consciously choosing to think with positive thoughts such as 'I express myself to the world' or 'I stand up for myself' or 'I will tell people what I like and what I don't like'. That's what I mean by embracing your potential because the only reason you *would* embrace potential is to urge positive thought. The only reason you would urge positive thought is to think positively such as 'I express myself to the world'.

NOTHING CAN HAPPEN THAT I CANNOT HANDLE

What's really important to remember is that your Soul only ever sends you challenges that are *slightly* higher than your current level of potential. This 'forces' you to search for the equally higher levels of potential, within you, to conquer and manage the challenges, but *never* at a level that's beyond that which you are capable of finding or re-membering. In that regard, *nothing* could happen to you today that you cannot handle. It's Cosmic Law.

The potential needed to conquer those challenges is already within the mind. All you would have to do is embrace the potential you already have and/or try to re-member the potential you have forgotten - with about as much effort as is required to remember someone's name or where you placed the fricken car keys! This means that any person, no matter what education or background, can conquer the challenges of life.

YOU also made sure that life surrounded you in potential to help re-mind you of your *own* potential. For example, Mother Nature is a living breathing example of pure potential. Teachers of potential (personal growth and spirituality) are everywhere and have been for centuries. Books like this one are appearing at an ever increasing rate by a diverse set of authors all over the world. The human race expects, rewards and worships potential. Children are the greatest exponents of joy (until their serious parents train them to be serious), which makes them the greatest teachers of potential. You naturally seek joy, which can only be created from powerful and wise potential. The shift in 2012 is the end of the old world of negativity and the beginning of a new world of potential. Life is built **on** potential. The Cosmos is built **from** potential. Potential is everything, everywhere and everywhen. You cannot fail.

NOTE - Anger is always personal and should NEVER be expressed towards others. Expressing anger safely is an art-form and The Yoooge Workshop contains some great tools and processes to cope with depression and express anger. - and it's free. For more details, go to www.yoooge.com and click on the 'Workshop' link.

WHAT IS GOOD AND WHAT IS BAD?

Life runs on a platform of extreme-duality and extreme-duality is a challenge in itself - one potential-aspect challenging its equal and opposite other potential-aspect. Every 'bad' challenge leads to a good outcome and every good challenge starts from a bad outcome. Otherwise, how would you know and why would you seek?

For example, the urge to be wealthy can only stem from a background of poverty. Success has historically been driven by the urge to overcome a problem. The need for creative expression is fuelled by depression. The need to know can only come from a standpoint of ignorance. The desire for love can only come from a standpoint of not-love.

In that regard, what is 'good' and what is 'bad'? In addition, challenges are so subjective. Any one of the following challenges in the table below might challenge one person in a 'good' way and another person in a 'bad' way.

For example, some people find the idea of learning new skills to be terrifying. Others find that same challenge to be quite rewarding. Yet the outcome is the same - an increased level of capability potential.

GREAT POTENTIAL CHALLENGES - GOOD OR BAD			
Security	**Capability**	**Expression**	**Knowing**
Bravery	Skills	Truth	Intuition
Poverty	Talent	Importance	Imagination
Risk taking	Achievement	Emotions	Understanding
Investment	Motivation	Purpose	Regret
Loss	Energy	Talent	Guilt
Greed	Success	Change	Worry
Competition	Criticism	Depression	Stress
Change	Failure	Suppression	Judgment
Deception	Completion	Conformity	Criticism
Danger	Goals	Powerlessness	Ignorance
Urges	Self-sabotage	Saying 'No'	Confusion
Adventure	Learning	Saying 'Yes'	Chaos
Fear	Frustration	Anger	Compassion
Anxiety	Anger	Embarrassment	Lifelessness
Mistrust	Rejection	Shyness	Boredom
Jealousy	Error	Creativity	Knowing
You	**You**	**You**	**You**
Relationships	**Relationships**	**Relationships**	**Relationships**
Parenting	**Parenting**	**Parenting**	**Parenting**
Life	**Life**	**Life**	**Life**

(Table) Any one of these great potential challenges might challenge one person in a 'good' way and another person in a 'bad' way. Yet the outcome is the same - an increased level of capability potential.

So there really is no good or bad; everything, everywhere and everywhen really is perfect. Once you really *get* this principle that nothing in life is 'bad', it opens the door to profound gratitude, which is one of the highest forms of OPTA and one of the fastest ways to en-Light-enment.

Gratitude is also one of the fastest ways to co-create yoooge experiences in your *you*niverse and create the life of your dreams.

THE FLOW-ON EFFECT

More importantly, humans are not qualified to judge life as either good or bad, so why bother? The necessary and default challenges of life and their flow-on effect are simply beyond the comprehension of humans. We couldn't judge life fairly, even if we wanted too. Your Soul and your Soul-Group are the only beings qualified to judge life *and* choose the default and necessary-challenges of life.

Your Soul is a God - an extremely powerful and supremely wise spiritual being that operates in the spirit-world, *outside* the limits of space-time. Your Soul can *simultaneously* view the effect of all your past lives, current lives and future lives - infinitely. So can your Soul-Group.

Your Soul knows exactly what level of potential is required for you in this lifetime, future lifetimes and even past lifetimes (your behaviour *now,* not only alters the outcome of future lives but also past lives). Your Soul therefore knows *exactly* what *kinds* of necessary-challenges you need in your life, to uncover the exact amount of potential(s) required to produce the perfect amount of default en-Light-enment, in every *moment*; in conjunction with the potential you will discover via your *own* purposeful-challenges.

That's a yoooge task. Even yoooger when you consider the flow-on effect of human actions. I like to use my story 'I Washed My Hands and Brought Down the Stock Market!'

to help demonstrate the massive flow-on effect of even the simplest action.

Imagine you're a travelling salesperson. You check into your motel after a long day. You've drunk too much coffee. You have trouble sleeping. That becomes *your* necessary-challenge. You get up at 2:30 a.m., go to the bathroom and then wash your hands. The sound of water squealing through the motel pipes wakes up the person sleeping in the motel room next to you. He gets very annoyed about being woken at 2:30 a.m. and has trouble getting back to sleep. That becomes *his* necessary-challenge.

He awakens in the morning tired and cranky and starts to have a *very bad day*. The negativity activates his EGO, which then feeds on itself with the bad, subconscious habit of reacting with negative thought and his mood becomes worse. He co-creates, attracts and activates events into his *you*niverse that match his mood. This makes him feel worse. His already depleted immune system crashes from his self-inflicted negativity and he catches a cold. He then passes that cold onto his daughter. That becomes *her* necessary-challenge

She goes to school and being seven years of age, plays with her snot, sneezes on her friends, sticks her rhinopharyngitis infected fingers in her friends' food and hugs a teacher. This spreads the cold around and becomes an entire school-community-based necessary-challenge. But let's focus on just *one branch* of that community necessary-challenge.

One of those children takes that cold home and coughs on her father. He gets a cold. He is always very busy at work because he has a high profile position as a stock trader. His stress weakens his immune system and his cold quickly turns to bronchitis and eventually into pneumonia. That becomes *his* necessary-challenge.

His doctor admits him to hospital where he's told to take an immediate break from work, to rest and recover. While resting, he totally forgets about a very important trade that's linked to a large company already on the verge of collapse. That forgotten trade wipes millions of dollars of value from that company's profile and it collapses. That affects seventeen other companies, linked to eighty-seven other companies. That becomes a *multi-company* necessary-challenge.

One hundred and fifty stocks are affected and over a few weeks, their prices plummet. That gets noticed, people panic and an already weak market declines. That declining market affects overseas markets and they decline. This sudden dramatic decline of stock prices, across a significant cross-section of the stock market, creates even more panic selling in a 'bear' market. This leads to double-digit percentage losses in stock market indexes over a period of several days. The stock market crashes. That becomes a world-wide necessary-challenge.

You simply washed your hands, yet you brought down the stock market. You had no idea that such a simple act would challenge so many people and in such a dramatic, yet necessary way. How could you? It's beyond human perception. Yet the flow-on effect of washing your hands challenged the entire human race (in some way) to re-member some form of potential. It's poetry in motion, or should I say, 'potential in motion'?

Your Soul and your Soul-Group can actually see the challenging flow-on effect of every thought-action-feeling *you* have ever made and will ever make and how that inter-connects with every other human-being on the planet. Now that's yoooge! We can't even fathom the flow-on effect of the simplest action. Imagine the *infinite* flow-on effect of all actions, all reactions, all responses, from all beings, from all time.

The default and necessary challenges of life are a GODlike decision, made by GODlike beings, with GODlike vision and precision. We humans aren't qualified to make those choices, understand those choices or comprehend those choices. We're certainly NOT qualified to judge those choices. Yet we expend *so much energy* and *waste so much time* doing exactly that. We're constantly judging, criticising and complaining about life, which also happens to be a major component of the EGO.

The easier path is to let go and go with the flow, knowing everything is perfect because it's chosen by our Soul, facilitated by our Soul-Group, within an illusion engineered by GOD and facilitated by Gods.

YOU DECIDE WHAT KIND OF CHALLENGE

Metaphysically speaking, space-time is an illusion and so there really is no space and there really is no time. Consequently, everything is happening *now.* Your Soul chooses all the default and necessary challenges in your life before you are even born. In that regard, your Soul ***pre***-chooses (***pre***) the challenges in your life. Your Soul also decides at what stage in your life to send you (***sent***) those challenges to you. This means that life is **pre-sent** by your Soul, as the pre-sent moment (the present moment).

You forgot the Truth so that you have no idea what pre-sent moment is coming next. The challenges of life are then pre-sented to you as a surprise. If *you* knew what was coming next then life wouldn't be such a challenge nor would you bother finding the potential required to overcome that challenge! You instead unwrap each moment like a present, as it comes to you, moment by moment. Life is a gift, that's why it's called the present (pre-sent).

More importantly, the challenge itself is more of a catalyst. Your actual behaviour (reaction-responses) to those challenges is what determines how you feel and *how you feel* becomes the real challenge. Cheering up from a bad mood is the real challenge. Maintaining a constant state of joy is the real challenge - especially when you're surrounded by so much negativity. That's why your Soul made an agreement between YOU and you that states:

'Your Soul chooses the default and necessary challenges in your life but you choose how you will *react or respond* when faced with those challenges'

Your reactions and responses determine how you feel. How you feel determines what type of challenge you will face. If you subconsciously react to life with negative thought (EGO), you feel bad, go downhill and create a difficult challenge. If you consciously respond well to life with positive thought (OPTA), you feel great, go uphill and create an enjoyable challenge. Even though the challenge may be a necessary-challenge, pre-sent by your Soul, you *still* get the choice to decide how difficult that challenge will ultimately end up being.

DÉJÀ VU

There is no space-time and everything is happening *now.* With that in mind, your Soul is pre-choosing your challenges in life *right now* and then sending them to you as pre-sent moments, *now.* Your Soul also knows exactly how *you* will behave when faced with that moment because that determines the challenge.

As a human, when that pre-chosen challenge arrives in your life, you behave *exactly* how *your Soul* knew you would. Your subconscious mind overlaps your Soul's subconscious mind. There is a moment, when the subconscious memory of your Soul choosing the challenge based on how you will behave, overlaps the memory of *you* experiencing that challenge and behaving, creating the sensation that *you have been here before,* otherwise known as déjà vu.

Your subconscious mind only overlaps your Soul's subconscious mind when you're in a relaxed or joyous state. This is why incidences of déjà vu mostly occur when you're holidaying or socialising.

WAVE TO YOUR SOUL

Did you voluntarily choose this book as a purposeful challenge or did this book just seem to coincidentally arrive in your life as part of a necessary challenge? The former was chosen by you and instigated by YOU. The latter was chosen by YOU and instigated by you. The potential effect is the same.

If this book was a purposeful-challenge, what inspired you to buy it? What's been happening in your life to urge you to seek out potential? If this book was a necessary-challenge, which incarnated-human-member of your Soul-Group brought it to you? How close to that person are you? What role do they play in your current life? Why did YOU (your Soul) pre-send this book to you?

This next bit is what we in the Society of Irreverent Metaphysical Philosophers, Larrikins and Entertainers (SIMPLE for short. I think I'm the only member so far), like to call a brain fart. A brain fart is when your brain tries to understand a really complex subject, yet the subject is just too mind-boggling. It holds onto the concept for a moment, contemplates, struggles, but eventually gives up and goes brrrsphthsphthsphth.

The concept that there is no space-time, especially when we rely on the reality of space-time, can cause brain farts. Consider this. YOU chose the challenge of reading this book before you were born. There is no space or time. That means that YOU are right *now,* choosing this book, as *you* are reading this line. If that's so, why don't you wave at YOU, who is *now* in the spirit-world choosing the event where you read the above line, and then waved at YOU, or didn't wave at YOU? Brrrsphthsphthsphth

Did you wave? Or did you mutter, 'I'm not waving at some stupid damn fricken … '. It doesn't matter. YOU already knew *exactly* how *you* would behave. That's *why* YOU pre-sented the above line to you, at this space and in this time. Brrrsphthsphthsphth

Freakier still, I have affected your life because you purchased my book. You have affected my life because you are reading my book. You and I must then be from the same Soul-Group. Awesome! Nice to meet you! That means that your Soul and my Soul are buddies and they both conspired together to get you to read this book. Even more brain-hurty is that your Soul and my Soul are *one,* and so *really,* YOU wrote this book about you! Brrrsphthsphthsphth

THE MEMORY-OF-POTENTIAL

Your Soul is a giant subconscious mind. It acts like a spiritual diary and keeps a cumulative record of the potential you have re-membered from each challenge. It stores the powerful and wise potential (the gold) *behind* every thought from every nonillionth moment of every past, present and future challenge - including your wishes, imagination and dreams.

It only stores the powerful and wise potential *behind your thoughts* - the gold - and it does so by filtering out any form of negativity. GOD is potential. In that regard, an ever cumulating memory of potential is an ever cumulating memory of GODlikeness - how GODlike have you become, lifetime after lifetime. This is what GOD observes. This is *how* GOD can observe all love from all beings from all time because She observes all memory-of-potentials from all beings from all time. She only observes the gold, the GODlikeness.

Your Soul uses your memory-of-potential to ensure the potential that *you* are born with in this lifetime, matches the maximum level of potential you reached in your *previous* lifetime. You then start your new life, onward and upward from there, re-membering higher and higher levels of potential from *that* potential point. You do this lifetime after lifetime.

You NEVER go backwards, ever. It's always onwards and upwards in ever-increasing levels of potential. You are here to rediscover your own Divine Magnificence in the same way that GOD is rediscovering Her own Divine Magnificence (it's actually one and the same thing). It's actually that gradually increasing, collective and cumulative human potential that expands the Cosmos at 75 km per second per 3.26 million years (The Hubble Constant).

In a way, your memory-of-potential 'drip feeds' potential to you, lifetime after lifetime, to deliberately limit your potential. That ensures the challenge of life, it makes evolution deliberate and slow, it prevents you from 'beginning at a level beyond you' and it also protects you from burning up in an electromagnetic flash.

YOU are a God with GODlike powerful and wise potential. But *you* can't have complete access to that potential - not as a human. Powerful and wise potential urges powerful and wise thought, which spiritually translates into Light, which then physically translates into light, heat and vibration. You as a human, trying to translate the full potential of your Soul into thought, would be the equivalent of trying to cram the Sun into a plastic Tupperware™ dish. It's just too bright, too hot and too loud.

Those of you who grew up in the 1980's may remember the scene in the 1981 adventure film 'Indiana Jones and the Raiders of the Lost Ark' when the bad guys open the lid of the Ark of the Covenant. The Spirit of the

Ark (god) then appears and melts the bad guys' faces off. A human experiencing the *full* potential of their Soul would be a bit like that except it's more of an electromagnetic, atomic, scrazoompity frrzzzerktaty kind of experience.

Your face would definitely melt off. The intense amplitude of the electromagnetic vibration would turn your biological flesh into a bloody mist, then vaporise that mist with an intense light of astonishing heat. The electromagnetic flash would kill most of the people around you but also wipe out every computer within a few thousand kilometres. Everything with a microprocessor would instantly fail, including your TV, car, phone, traffic lights, aeroplanes and more. You can see how that much Light wouldn't work in the illusion, as a human-being.

Mother Earth has the same problem. She re-members *all* of Her potential but She converts the scrazoompity frrzzzerktaty energy of Her potential into heat and light. That heat is hot enough to melt rock and metal into a fluid called magma (lava), which is why She has a molten core. The iron in the Earth's core helps Her express Her magnetic-potential as the Earth's magnetic field. The minerals within magma help Her express electrical-potential as static electricity and lightning. Those minerals and metals form into crystals and still contain the electromagnetic fingerprint of Mother Earth's potential. That's why the electromagnetic vibration of crystals can heal the body and calm the mind.

The Sun also converts its potential into heat and light, as do the stars and galaxies. Other Cosmic elements that cannot translate potential into heat and light recycle their potential. They translate the potential in reverse, creating huge vacuums in space-time known as Black Holes (just a theory), which are powerful enough to suck away unnecessary translated light, heat, vibration and gravity from other beings as well.

INTUITION

YOUR memory-of-potential forms part of *your* upper-aspects of your subconscious mind and so *you* can tap into this memory-of-potential anytime you're in a relaxed state of joy. When you do so, the potential *within that memory* influences your brain to think powerful and wise thoughts, based on many lifetimes of re-membered potential. This influence is otherwise known as intuition.

Your intuition is the 'voice in your head' or your 'gut feeling' that keeps you out of danger, helps you make better decisions, helps you be more creative and connects you with YOU. If you delve deeper into your intuition you find it to be a block of thought accompanied by a *feeling* to create a *knowing*. This is where we get the intuitive phrases such as, 'I just *knew* it was the right thing to do' or 'It just didn't *feel* right to me'.

The *knowing* part of your intuition is your brain being influenced by the potential within YOUR memory-of-potential. The *feeling* part of your intuition is the emotional result of your intuitive thoughts. They combine to create a block of thought, accompanied by a feeling, to create a knowing.

Your intuition reminds you how far you've come, potentially, from previous lifetimes. It stops you from going *backwards* in potential. Your intuition is often very subtle because the feelings generated by intuition usually match your current level of potential. It's only when you do something considerably less than your current potential, otherwise referred to as weak (dangerous, harmful to others, self-destructive, incompetent) or foolish (rude, untrue, ignorant, stupid) that your intuition becomes really noticeable.

If you do something 'less than your potential', or for want of a better word, 'stupid'; the thoughts of weakness/foolishness behind that behaviour will noticeably clash with your intuition to immediately inform you that what you are doing is *really stupid*. This creates the all too common, 'I just knew/felt I shouldn't have done that' kind of knowing-feeling.

Intuition is subtle for a reason. *You're* supposed to make your own choices based on *your* security, capability, expression and knowing. That's part of the challenge of life. If you can't make a choice, it's because you lack the potential to make that choice. Finding that potential is why you're here. Your Soul can't interfere with that process because it nullifies the challenge. It can *slightly influence* that process with intuition, but it can't tell you what to do.

You have to decide what to do in life and your intuition will either agree with you or it will offer you a bad knowing-feeling that building a ramp and trying to triple-somersault a snow plough over a school bus, 'high on crack' is really stupid. Otherwise, choice itself *is* the challenge.

YOU KNEW THE JOURNEY WOULD BE SLOW

Whether you choose a purposeful-challenge, or whether *YOU* chooses a necessary-challenge *for* you, the EGO deliberately keeps you locked in a negativity feedback loop. Cheering up from the EGO is the challenge.

You go round and round on the cycle of life, same negative thought patterns over and over again, same minor journeys into impurity, over and over again. Each time you return from a minor journey into impurity, you uncover a slightly higher level of potential, making you slightly wiser and more powerful *each time*. Guess what happens if you go round and round, each time slightly higher? You create a spiral.

It's not surprising that the spiral is a sacred symbol. The spiral is represented in nature in seashells, cyclone cloud formations, patterns in plants, the DNA helix and rotating galaxies. The sacred mathematics of the Mandelbrot Sets and the Fibonacci numbers (0, 1, 1, 2, 3, 5, 8, 13, 21, 24, 55, 89, 144 ...) both create spirals. The spiral is represented in Celtic patterns, indigenous artwork and even modern architecture. It's also no surprise that the journey of en-Light-ment also follows the path of a spiral.

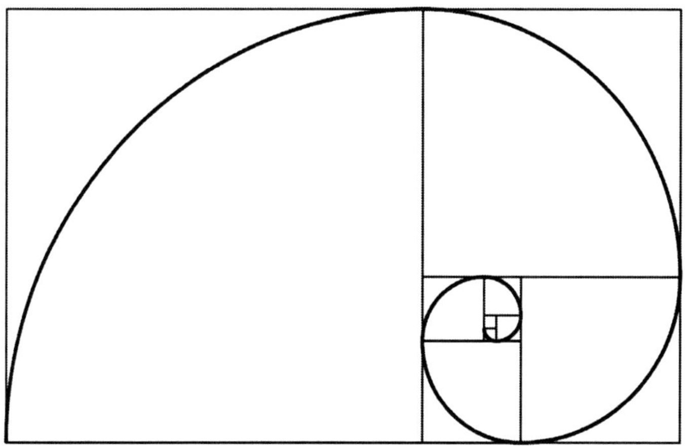

(Figure) A spiral created by drawing circular arcs connecting the opposite corners of squares - squares whose sides are successive Fibonacci numbers in length, sized 1, 1, 2, 3, 5, 8, 13, 21, and 34. Notice how the spiral arcs parabolically (not evenly) and returns to the centre?

A true spiral is not even. It winds outwards from the centre parabolically, like a snail's shell (see figure above). Just like the snail-shell-spiral, we humans start the journey of en-Light-ment at the centre and then lifetime after lifetime we work our way outwards towards en-Light-enment (purity of thought).

The spiral of en-Light-enment is three dimensional because we humans go round-and-round with EGO as we move onward and *upward* with potential. A three-dimensional spiral not only expands and ascends, but the journey on the outer edges of the spiral gets faster and faster, as it expands.

The journey of en-Light-enment *starts out* slow like a snail. A slow journey is more deliberate, certain, learned and permanent. A fast journey misses the point and many of the lessons. It then parabolically moves away from the centre because life has a parabolic increase in joy as you gain more potential. The journey of en-Light-ment gets faster and faster with age and with each lifetime - except like a spiral, it never ends and always expands into higher and higher levels of GODlikeness. Woot! Woot!

SUMMARY

- Anything in life that is challenging - good or bad - will 'force' you to find the potential within yourself to facilitate, master or conquer that challenge.
- This helps you re-member potential.
- The Gods set up life to be challenging.
- YOU got together with other Gods to form a Soul-Group.
- You incarnate with human-members of your Soul-Group as family, friends, colleagues and various strangers.
- The job of the Soul-Group is to help set up *your life* to be challenging.
- The incarnated-human-members of your Soul-Group then challenge each other on various different levels to ensure everyone in that Soul-Group evolves.
- You are expected to voluntarily challenge yourself with purposeful-challenges.
- If you don't, YOU will challenge you, for you, with necessary-challenges.
- The potentials required to overcome the necessary-challenges are often the same potentials that you are choosing to ignore.

- It doesn't even matter if you don't conquer the necessary-challenge because the real challenge is how you react or respond to that challenge.
- If you subconsciously react to life with negative thought (EGO), you feel bad, go downhill and create a difficult challenge.
- If you consciously respond well to life with positive thought (OPTA), you feel great, go uphill and create an enjoyable challenge.
- Cheering up from a bad mood or maintaining a constant state of joy *is* the real challenge.
- There is no 'good' or 'bad' and everything is perfect.
- Humans aren't even qualified to judge whether life is good or bad.
- The default and necessary challenges of life are a GODlike decision, made by GODlike beings, with GODlike vision and precision. It's none of our business and easier to assume that everything is perfect.
- Your Soul chooses the default and necessary challenges in your life but you choose how you behave when faced with those challenges. That, again, is the real challenge.
- The concept of space-time can cause brain farts.
- Your Soul keeps a cumulative record of the potential you have re-membered from each challenge.
- It only stores the powerful and wise potential (the gold) *behind* every thought as a memory-of-potential.
- Your memory-of-potential exists in the upper reaches of your subconscious mind and feeds your intuition.
- Intuition is often subtle until you do something really stupid.

- Then you get a block of thought accompanied by a feeling to create a knowing/feeling that you are being really stupid.
- Your memory-of-potential drip feeds potential to you, lifetime after lifetime.
- It's always onwards and upwards.
- You NEVER go backwards, potentially.
- The journey is slow like a snail but parabolically gets faster.

Challenge yourself in a purposeful way or
life will challenge you in a necessary way.

POSITIVE THINKING

Throughout this book I have mentioned the concept of consciously choosing to respond to life with positive thought, with the end goal being *only* positive thought-action, hence the acronym of OPTA. This chapter shows you exactly what positive thoughts to choose, how they are formed and why humans *positively* think the way they do.

The two major potentials of power and wisdom break down further to each have a lower and upper aspect, making four potential-aspects in total. Those four potential-aspects influence the brain to think in four distinctly different, *positive* ways.

Please study these four ways of positive thinking at the end of this chapter. If that's the only thing you get out of this book, then fantastic. If you can consciously choose these types of thoughts, no matter what happens around you, your life will change, for the better, at a rate that is nothing short of miraculous.

YOU CO-CREATED YOU

Your Soul and your Soul-Group set up the challenges in your life. Your Soul then does the most extraordinary thing. It co-creates you in the same way the Gods co-created other Gods - ONE-TWO-THREE-10,000-things.

Your Soul thought of the human-you. The *thought-of-the-human-being* was full of powerful and wise potential. The potential within the *thought-of-the-human-being,* urged the *thought-of-the-human-being* to think. A thinking being is a conscious, living being so when the *thought-of-the-human-being* thought, it came alive. It became the ONE.

The first thing the ONE thought about was itself. *It* was powerful and wise potential, so when it thought of itself it contemplated its own power and wisdom, which is to say it contemplated the potential of its 'parent', your Soul (YOU).

To do so, and avoid the potentials behind the contemplation cancelling out to nothingness, the ONE split into two minds. It became the TWO. The TWO then pulsed between its two minds, to separately embrace its two potentials, to think the powerful and wise thoughts necessary to contemplate its Soulful potential.

The pulse of that contemplation created ripples of electromagnetic vibration in the Cosmic-pond. The TWO became THREE (vibration).

The ripples of the THREE then interacted with the ripples of all other beings and the ripples of space-time to co-create some*thing*, some*where*, some*when*. The THREE became 10,000 things - the human-being.

The 10,000 things of the human-being is a vibrational-reality that extends from the human body to the human energy field (aura) to the astral-world (mind), to the spirit-world, to the *thought-of-the-human-being* and finally to the point of Light that thought the *thought-of-the-human-being,* which is YOU.

The vibrational-reality of the human-being-you is co-created in the image and likeness of the Soul (YOU). You have the same access to the exact same potential of your Soul (eventually). The only difference between you and YOU, is that you forgot the Truth. You forgot you're a God, connected to all things living with a giant illusion and so you believe you're a human-being, separate from all things, living within reality. You also forgot most of your Soulful potential.

Or at least that's the best way I can describe what is essentially: different frequencies of the same vibrational-reality existing as different elements that *appear* separate within an illusion but are actually all connected, as *one*.

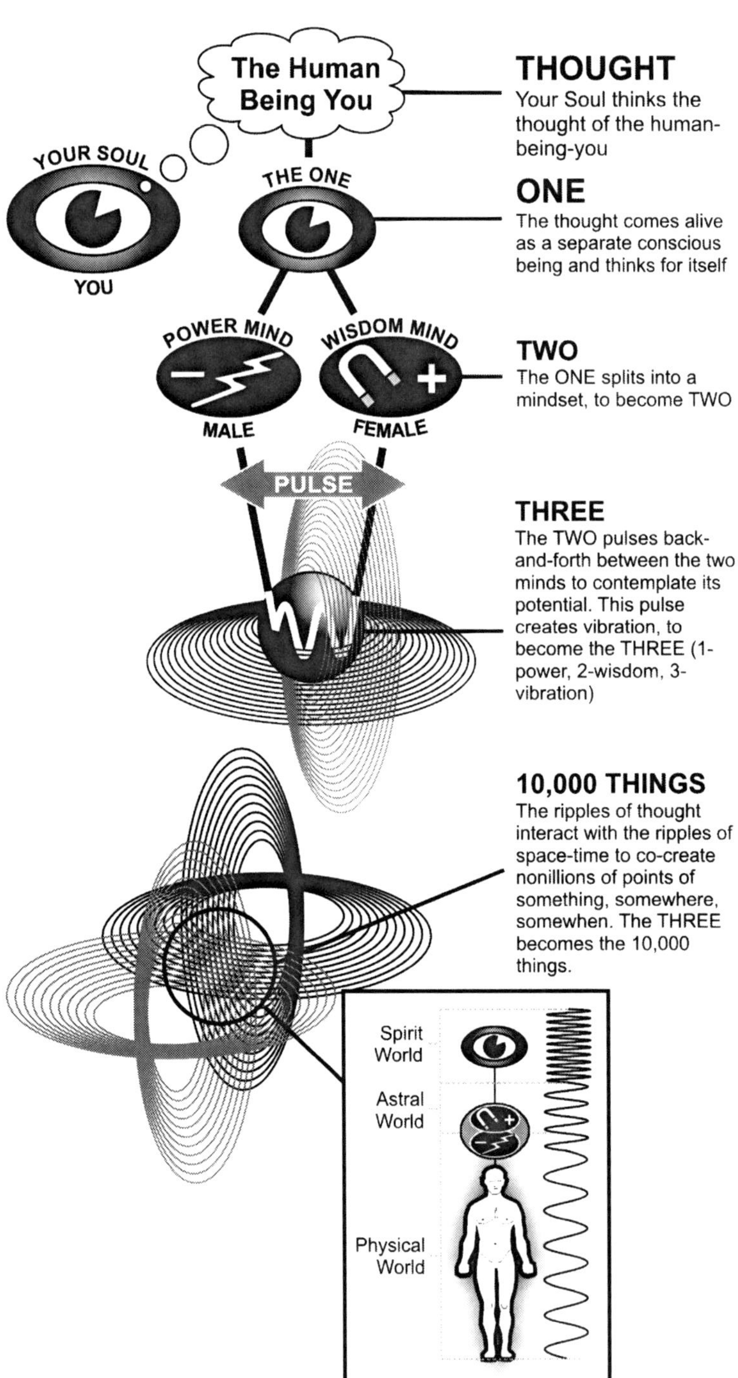
The Human Being You
YOUR SOUL
YOU
THE ONE
POWER MIND
WISDOM MIND
MALE
FEMALE
PULSE
THOUGHT
Your Soul thinks the thought of the human-being-you
ONE
The thought comes alive as a separate conscious being and thinks for itself
TWO
The ONE splits into a mindset, to become TWO
THREE
The TWO pulses back-and-forth between the two minds to contemplate its potential. This pulse creates vibration, to become the THREE (1-power, 2-wisdom, 3-vibration)
10,000 THINGS
The ripples of thought interact with the ripples of space-time to co-create nonillions of points of something, somewhere, somewhen. The THREE becomes the 10,000 things.
Spirit World
Astral World
Physical World

1. **THE ONE** - The potential within the *thought-of-the-human-being* urges it to think and it comes alive as a *separate* conscious being and it thinks for itself. It becomes the ONE.
2. **THE TWO** - The ONE splits into two minds, to form the mindset of power and wisdom. The ONE becomes the TWO.
3. **THE THREE** -The TWO pulses back-and-forth between the two minds to contemplate its potential. This pulse creates ripples in the Cosmic-pond - vibration. The TWO becomes THREE.
4. **THE 10,000 THINGS** - The ripples interact with the thought-ripples of all other beings and space-time to co-create nonillions of points of vibrational-reality - some*thing*, some*where*, some*when*. This becomes the form, function and phenomena of the human-being that extends from the human body, to the human aura, to the astral-world and to the spirit-world.

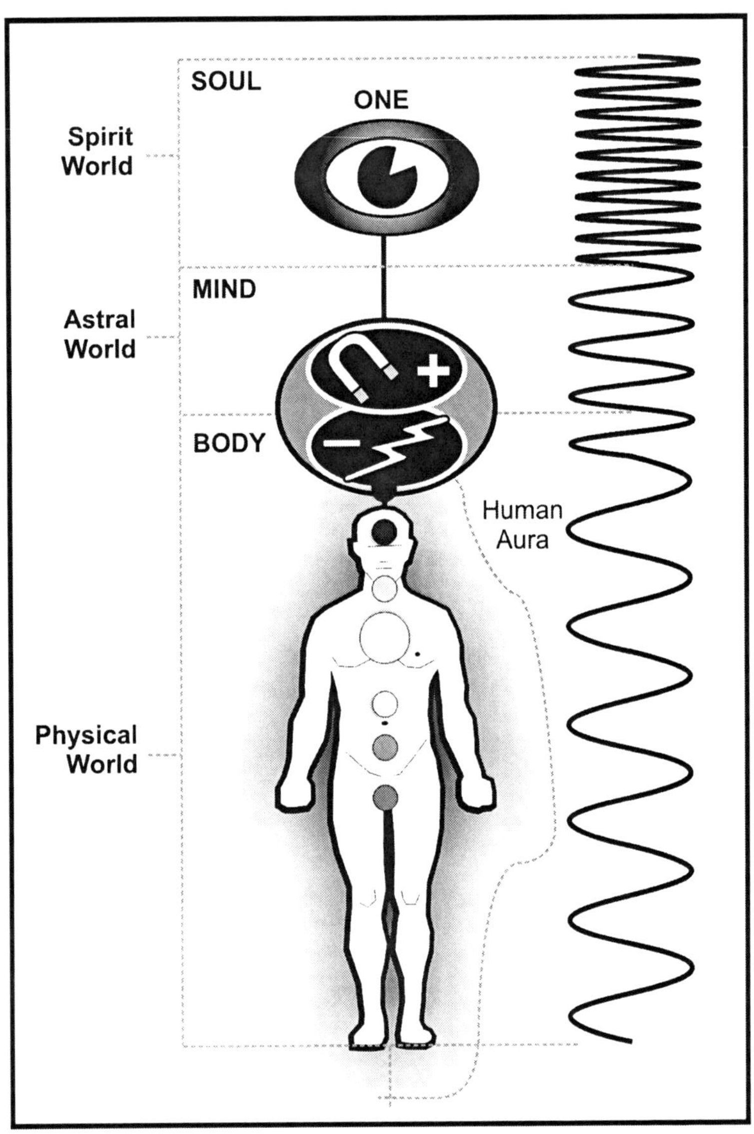

(Figure) The vibrational-reality of a human-being extending from the physical-world to the astral-world to the spirit-world.

THE CONTEMPLATION LEVEL AND HUMAN LEVEL

In essence, the fundamental source of the human-being is the mindset (the TWO). That's what creates the pulse that produces the vibrational-reality of the human-being that extends from the human body to the aura to the astral-world to the spirit-world.

That mindset urges thought on two distinctly different levels. It first contemplates its own power and wisdom on a *contemplation-level* (contemplation-mindset). That pulse creates the vibrational-reality of the human-being that extends from the human body to the aura to the astral-world to the spirit-world.

The same mindset then urges the human brain, within the human body, to think powerful and wise thoughts on a *human level* (human-mindset) and at a level predetermined by how much potential you have re-membered, because you forgot the Truth.

That means the mindset simultaneously thinks on two different levels - a contemplation level (contemplation-mindset) and a human level (human-mindset). It can do this easily because the human-being is a living-thought (ONE), thinking on a contemplation level (TWO), to co-create the vibrational-reality of itself (THREE), as a human-being (10,000 THINGS) that extends from the human body to the aura to the astral-world to the spirit-world. Brrrsphthsphthsphth.

Vibrational-reality vibrates in and out of nothingness, nonillions of times a second. It all comes from nothingness and it all returns to nothingness and so vibrational-reality can pretty much do whatever it wants. It can *simultaneously* be this, or that, or both or neither, all at the same time and in the same space. It can easily think on two levels.

The contemplation-mindset pulses back-and-forth between the physical-world and the spirit-world to create the vibrational-reality of the human-being-you that extends from the physical-world to the spirit-world. The contemplation-mindset always contemplates itself in the background and that contemplation will always create the vibrational-reality of the human-being, even long after you have died (your dead body, the after-death aura, the vibrational-reality of the rotting body, your legacy, etc).

The human-mindset pulses at a much lower level, back-and-forth between the physical-world and the astral-world to co-create events in the *you*niverse. This pulse creates ripples in the Cosmic-pond that interact with the ripples of space-time and the ripples of all other beings, to co-create *events* in your *you*niverse. Your human-mindset thinks when you want to. It can embrace the powerful and wise within the human-mindset and respond to life with positive thought or it can ignore the powerful and wise potential within the human-mindset, creating an opportunity for the EGO to react to life with bad habits of negative thought.

THE HUMAN LEVEL

The spirit-world in its purity cannot tolerate the impurity of human thought, especially the subconscious mind which contains a lifetime memory of impure thought. Consequently, the human-mindset exists in a dimension that's not quite physical and not quite spiritual but *astral*. The astral-world is a special layer of spirit-world that can tolerate the impurity of human thought. It could be considered the lowest vibration of the spirit-world and it acts like a bridge between the physical-world and the spirit-world.

The astral-world is very often *called* the spirit-world but mainly because ghosts hang out there and so it's the *spirit*s world. Technically speaking, the astral-world is a low-vibrational layer of the spirit-world, which can accommodate the lower-vibrational energy of the human thought. The astral-world is also referred to as the fourth dimension and the dream world.

When you think, you embrace your potential, which then urges the brain to think. The mindset then pulses back and forth between the physical-world and the astral-world to accommodate each mind within the mindset. The power-mind can only operate within the framework of space, in the physical-world. The wisdom-mind can only operate within the framework of time, in the astral-world. When the human-being thinks it pulses into the physical-world, thinks powerful thoughts; then pulses into the astral-world, thinks wise thoughts; then pulses back to the physical-world and so on. It does this over and over again, nonillions of times a second; physical-world, astral-world, physical-world, astral-world, which creates a pulse.

The human-mindset doesn't come pre-loaded with potential. It starts out blank (zero-point). The physical-world and the astral-world are *loaded* with potential since they are made of potential. The physical-world is full of powerful-potential and the astral-world is full of wise-potential. As your human-mindset pulses in-and-out of the physical-world and the astral-world, *both minds* come with you for the ride. They have to. You're a mindset.

Each mind then soaks up the potential within each world. In other words, the power-mind soaks up the powerful-potential in the physical-world and the wise-potential in the astral-world. The wisdom-mind soaks up the powerful-potential in the physical-world and the wise-potential in the astral-world.

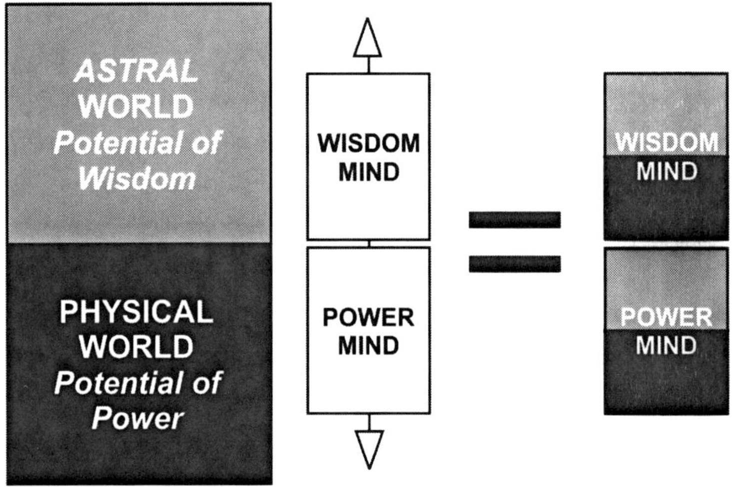

(Figure) The two minds soaking up the potential of power and the potential of wisdom as they are rapidly dipped in-and-out of the physical-world and the astral-world.

Because you forgot the Truth and now think on a human-level, you can only soak up potential at a human level. Your mindset pulses at a frequency that matches the potential stored in your memory-of-potential and so it can only soak up potential within that level (that's actually why your mindset doesn't pulse past the level of the astral-world).

Actually, the level of potential available to soak up is always *slightly* higher than your memory-of-potential. This ensures that every new thought can be urged by a potential that's above your current level of re-membered potential, in order to facilitate, conquer or master a challenge that is often slightly above your potential.

Whether or not you embrace or ignore that potential is still up to you but that potential is always there, always fresh and always slightly above you. This 'forces' you to embrace higher levels of potential. It's always onwards and upwards. If you choose to, you can think positively at any moment. It may take some effort, especially if you're in a really bad mood, but it's possible. That's why all humans cheer up - eventually.

SOAKING UP POTENTIAL

Energetically speaking, your power-mind runs on its own energetic framework of space and your wisdom-mind runs on its own energetic framework of time.

When you think, your human-*mindset* pulses into the physical-world. This immerses the human-mindset (both minds) in the potential of power. The power-mind soaks up that potential-of-power in the physical-world but it does so over its own framework of space. That adds the potential-aspect of **power-over-space** to the power-mind. Meanwhile the wisdom-mind soaks up the *same* potential-of-power in the physical-world but it does so over its own framework of time. That adds the potential-aspect of **power-over-time** to the wisdom-mind.

Once you have finished with your nonillionth-of-a-second pulse in the physical-world, you pulse over into the astral-world. The human-mindset (both minds) is immersed in the potential of wisdom. The power-mind soaks up the potential-of-wisdom in the astral-world but it does so over its own framework of space. That adds the potential-aspect of **wisdom-over-space** to the power-mind. Meanwhile the wisdom-mind soaks up the *same* potential-of-wisdom in the astral-world but it does so over its own framework of time. That adds the potential-aspect of **wisdom-over-time** to the wisdom-mind. This creates four distinctly different potential-aspects.

1. **POWER-MIND 1** - **Power** *Over* Space
2. **WISDOM-MIND 2** - **Power** *Over* Time
3. **POWER-MIND 3** - **Wisdom** *Over* Space
4. **WISDOM-MIND 4** - **Wisdom** *Over* Time

When you line them up to be mind specific, the power-mind becomes the two potential-aspects of power-over-space and wisdom-over-space. The wisdom-mind then becomes the two potential-aspects of power-over-time and wisdom-over-time. These four potential-aspects influence the brain to think in four different ways.

1. **POWER-MIND 1** - **Power** *Over* Space
2. **POWER-MIND 3** - **Wisdom** *Over* Space
3. **WISDOM-MIND 2** - **Power** *Over* Time
4. **WISDOM-MIND 4** - **Wisdom** *Over* Time

The reason why the pulsing-mindset soaks up potential in this order is simple - it stops the potential-aspects from cancelling each other out to zero. The potential-aspects of power-over-space and power-over-time are both lower-aspects (potentially) and so are both -ve. Those potential-aspects can co-exist together without their potentials cancelling each other out to nothingness. The same applies to wisdom-over-space and wisdom-over-time. They're both upper +ve aspects and so can happily co-exist without nothingness.

But those potential-aspects are Cosmic. We're not. We're human and we live in the 'real' world (whatever the %$#@ that is). We're surrounded by the four major elements of life - ourselves, other people, the world and things - and those real-world forces heavily influence the thoughts we think. We don't think positively like Gods, we think positively like humans.

When you combine the above four potential-aspects with the four main elements of life, you get four main ways that humans think positively. For example:

THE FOUR POTENTIAL-ASPECTS OF THE MINDSET

1. **POWER-MIND** - **Power** *Over* Space
2. **POWER-MIND - Wisdom** *Over* Space
3. **WISDOM-MIND - Power** *Over* Time
4. **WISDOM-MIND - Wisdom** *Over* Time

THE FOUR MAJOR ELEMENTS OF LIFE

1. **YOU** - body (health, sexuality, skills), mind (belief systems, thought, imagination) and soul (emotion, desire, inspiration)
2. **PEOPLE** - family, relationships, friendships and strangers
3. **YOUR WORLD** - fulfilment, career, money, adventure and life experience
4. **THINGS** - necessities, possessions, wants and needs, comfort and money.

THE FOUR THOUGHT SETS

1. **Power** *Over* Space + Life = **SECURITY**
2. **Wisdom** *Over* Space + Life = **CAPABILITY**
3. **Power** *Over* Time + Life = **EXPRESSION**
4. **Wisdom** *Over* Time + Life = **KNOWING**

SECURITY (LOWER POWER)

The potential-aspect of *power over space* means 'on-off' control over space. As a human, 'space' is physical-world reality (you, people, your world and things), and so an on-off control over space is more like a 'Yes-No' control over yourself, other people, your world and things. That's more of a fight-or-flight-survival come supply-and-demand type of influence. This potential-aspect-life-combination would eventually urge the brain to think:

- **I am brave and strong.**
- **I trust people; they are supportive and kind.**
- **My world is a safe and wonderful place.**
- **All my wants and needs are supplied by the unlimited creative power of my Soul. There is no limit. I can do, be, or have whatever I desire.**
- **I AM SECURE.**

CAPABILITY (UPPER POWER)

The potential-aspect of *wisdom over space* means 'push-pull' control over space. As a human, push-pull control translates as *choice*. Space is physical-world reality. Therefore, push-pull control over space is more about choices, involving you, people, your world and things.

This differs from the more urgent, needy, supply-and-demand properties of security because capability has a more relaxed element of choice. It's the 'want' rather than the 'need'.

One potential aspect in a mind feeds the other. The potential-aspect of security (courage and abundance) feeds the potential-aspect of capability, to create thoughts such as 'I am secure, so *I can* make these choices. Therefore, I can do it'. This potential-aspect-life-combination would eventually urge the brain to think:

- **I am an amazing and talented person who is capable of anything and worthy of everything. I have the strength, willpower and skills to achieve my goals to do, be, have whatever I desire.**
- **I choose people in my life that help, guide and support me, to be what I choose to be. I am appreciated by everyone in every way.**
- **The world supports my success.**
- **Everything I need to achieve my goals will be supplied by the unlimited creative power of my Soul.**
- **I AM CAPABLE.**

EXPRESSION (LOWER WISDOM)

The potential-aspect of *power over time* means 'on-off' control over time. As a human, '*over time*' is memory, so on-off control over time means on-off control of memory. Your memory is the subconscious collection of all your past experiences. It's what makes you, you. Switching that on-off would translate to instant expressions of you: physically, verbally, emotionally and mentally - your personality. This potential-aspect-life-combination would eventually urge the brain to think:

- **This is me. I am unique, I am important and I matter. I easily *express* myself, my emotions and my point of view.**
- **My ideas, point of view and boundaries are valid and will be respected by all. I speak up to create and embrace change in my life. I say 'Yes' and I can say 'No'.**
- **I share my inspiration and desires with the world.**
- **I am artistic, creative and intelligent.**
- **I EXPRESS MYSELF.**

KNOWING (UPPER WISDOM)

The potential-aspect of *wisdom over time* means push-pull control over time. As a human, push-pull is choice and *over time* is memory. Therefore push-pull control over time is choice of memory and that translates to knowledge, imagination and life-experience. It's what feeds your knowing (not just knowledge but knowledge, life-experience and imagination combined). This potential-aspect-life-combination would eventually urge the brain to think:

- **I know that I am secure, capable, expressive and knowing.**
- **I know how to have fantastic, trusting, loving relationships with people.**
- **I know my life's purpose and my higher purpose. I know the next step. I trust my intuition and make perfect decisions every time. I know I am guided by my Soul.**

- **I know what I like and what I don't like. I know what to do, be and have.**
- **I KNOW.**

SUMMARY

- YOU created you as a human-being in the same way the Gods co-created other Gods - ONE-TWO-THREE-10,000-things.
- The human-being-you is the 10,000 things that extend from the human body to the aura to the astral-world to the spirit-world.
- You are the human-body part of the 10,000 things.
- The fundamental source of the human-being is the mindset (TWO).
- The mindset thinks on a contemplation-level and a human-level.
- On a contemplation-level, the pulse of the mindset creates the vibrational-reality of the human-being that extends from the human body to the aura to the astral-world to the spirit-world.
- The contemplation-mindset always contemplates itself in the background and that pulse maintains the form, function and phenomena of the human-being.
- The human-mindset pulses at a much lower level, back-and-forth between the physical-world and the astral-world.
- The human-level pulse co-creates events in your *you*niverse.
- On a human-level the mindset urges the brain to think powerful and wise thoughts.
- Your human-mindset thinks whenever it wants to.
- The human-mindset doesn't come pre-loaded with potential.

- It soaks up the potential within the physical-world and the astral-world as it pulses back-and-forth between the power-mind and the wisdom-mind.
- It pulses at a level that matches the memory-of-potential and so it soaks up potential at that level or slightly higher than that level.
- The power-mind soaks up the powerful-potential in the physical-world over a *framework of space* and the wise-potential in the astral-world over a *framework of space.*
- The power-mind soaks up that potential-of-power in the physical-world over a framework of space to create the potential-aspect of power-over-space, which is **security.**
- The wisdom-mind soaks up the potential-of-power in the physical-world over a framework of time to create the potential-aspect of power-over-time, which is **expression.**
- The power-mind soaks up the potential-of-wisdom in the astral-world over a framework of space to create the potential-aspect of wisdom-over-space, which is **capability.**
- The wisdom-mind soaks up the potential-of-wisdom in the astral-world over a framework of time to create the potential-aspect of wisdom-over-time, which is **knowing.**
- This creates four distinctly different potential-aspects.
- **SECURITY** - I am safe and secure, I trust people and my world is a safe and wonderful place.
- **CAPABILITY** - I can do it, I can be successful, I can achieve my goals and I am worthy of success.

- **EXPRESSION** - I can express myself, my point of view, my boundaries and my creative-intellect to the world.
- **KNOWING** - I know myself (know thyself), people, my life's purpose and what I like.

OPTA

The four preceding thought-sets of security, capability, expression and knowing are a perfect example of how to think with positive thought. They're not just affirmations. They're extremely powerful ways of positive thinking because they specifically target (and hence overpower) the negative thoughts that make up your EGO.

Thinking these types of positive thoughts, as much as you can, no matter what is happening around you, is how you develop a good habit of consciously choosing positive thought, which is OPTA.

OPTA

OPTA stands for Only-Positive-Thought-Action. 'Action' gets a mention because your thoughts drive your actions. You can think about security, capability, expression and knowing all you like but eventually you have to get off your arse and *do* something about it, by taking positive action. Positive action not only physically expresses positive thought but positive action also inspires positive-thought.

The 'Only' in 'Only-Positive-Thought-Action' refers to the eventual end goal, where positive-thought-action has become such a natural part of you that it's the *only* way you think-act-feel. That's when you become OPTA-mised. It's a Jesus-like state that takes many lifetimes to master.

The end goal may take lifetimes to master but the practice of OPTA is simple. It's not easy but it *is* simple. The first step of OPTA is to ban negativity from your thinking. No negativity whatsoever. No complaining, criticising, worrying, gossiping, aggression or misery. Above all, try and avoid the four common types of negative thought - insecurity, failure, depression and ignorance (see table on page 168). Avoid people who do any of the above. Steer clear of media that expresses any of the above such as 'The Nightly News', violent movies, TV drama, or most of the Internet (especially conspiracy theories).

The second step of OPTA is to practise the four preceding thought-sets of security, capability, expression and knowing, as much as you can. Get to know them. Print them out and stick them on the shower screen (the outside, facing in), the toilet door, your office wall and your bedside. Repeat them to yourself first thing in the morning and last thing before you go to bed. It takes less than two minutes to read out loud. Write the words OPTA on pieces of paper and stick them everywhere to remind you to choose positive thought. Practise OPTA fanatically for at least twenty-eight days - the time it takes to form a new habit.

Most importantly, you can't be self-critical, blame yourself or feel guilty if you fail at your OPTA attempts (or OOPTA as we call it). You're only human and the negativity of the EGO is a natural part of your thinking. If you make a mistake and think negatively, then so be it. The practice of OPTA is a gentle, self-worthy practice. Not an excuse to beat yourself up because of your negativity. The challenge is to gently resolve the EGO, as you practise

OPTA. The most important thing to remember when practising OPTA is:

The practice of OPTA is impossible when you're in a bad mood.

It is impossible to practise OPTA when you're in a bad mood because OPTA creates the feeling of joy (eventually). If you're in a bad mood, the gap between the feeling of joy and the feeling of your bad mood is just too great. It's too much of a leap. You'll never succeed at OPTA if you're in a bad mood.

You have to cheer up first *then* practise OPTA. OPTA was designed to amplify a good mood into happiness and then into joy. This re-acquaints a person with the feeling of joy (a feeling that most adults have long forgotten) and helps develop a habit of consciously choosing positive thought. It was never designed to cheer people up. Cheering up from a bad mood is the challenge. Once you master that challenge, then you can practise OPTA.

In a good mood, OPTA is more believable. There is less chance your EGO can be activated and fight you with a 'Yeah right!' response to OPTA thought-sets. You don't even have to fully believe what you read. In a good mood, the subconscious mind does not question whether your thoughts are real, unreal, fake or imaginary. It's passive anyway and simply stores and categorises the information into habits of thought. In a good mood, it stores the good-mood-thoughts of OPTA in the good mood section, creating a powerful subconscious storehouse of positive thinking that erodes the EGO.

CHEERING UP

Cheering up from a bad mood is a challenge. It's actually *that challenge* that 'forces' you to search for the potential within yourself, to urge thoughts positive enough, to create the feel-good-feelings required, to cheer you up from your bad mood. If you bear that in mind then anything in life that makes you feel bad is actually a potential gift, often pre-sented by YOU, to accelerate your en-Light-enment, and often because *you* don't purposefully-challenge yourself enough. Ponder that and you'll rarely feel bad for very long!

Remember that *you* put *yourself* in a bad mood. Your thoughts create your feelings and so how *you* react to life is what makes you feel bad. Whether you got angry or frustrated, whether you did something wrong or failed, whether someone said something hurtful to you, or whether you were treated unfairly; it was *your* thoughts behind how *you reacted* to that situation that put you in a bad mood. Trying to cheer up from that bad mood by changing those *same* thoughts into more positive ones is unlikely, and often impossible.

For example, if someone does something that makes you angry, *they* didn't make you angry. You subconsciously reacted to that situation *badly,* with angry thoughts about that person and those angry thoughts about that person made you feel anger. The other person may well be an inconsiderate buffoon, but you hurt your own feelings, with your own bad reactions. You then go downhill into a bad mood.

Cheering up from that bad mood by changing your thoughts about the buffoon is nigh impossible. Your EGO has you trapped in a negativity feedback loop of anger. Your angry thoughts are making you feel angry which then activates even more angry thoughts. Trying to bust out of that loop by changing those very thoughts is too difficult. You are too attached to those thoughts and the anger associated with the buffoon. It's too raw, too soon and too hard.

You *can try* but most of your EGOic imagination will probably be preoccupied with reliving the argument in your head, saying all the things to the buffoon that you never thought of, had the guts to say or remembered to say at the time. Your EGOic imagination may even be totally preoccupied with the mental image of rapidly rubbing the buffoon's face, back and forth across the pavement to give them a nasty 'Chinese Burn' - concrete style. Maybe not, but you get the point.

To cheer up, acknowledge the anger. Ignoring your anger or suppressing it will only make you angrier. Your feelings are valid - all of them, including anger. That's why it's there, sent by YOU, as a necessary-challenge. Acknowledge that you are angry and that it's OK to be angry. Then get angry. *Feel* the necessary-challenge. Yell, spit, scream or stomp on the ground. My personal favourite is screaming into a pillow or underwater. I also like to break stuff and occasionally I will punch a cardboard box! Just remember: anger is personal and should NEVER be directed at others. There are many NASA types who will tell you that it's not a good idea to get angry, that anger activates neuro-peptides and creates more anger and we should instead transmute our anger with deep breathing and meditation and ... %$#@ that! Anger is a valid emotion, a necessary-challenge and needs to be expressed. It's only when it's NOT expressed that it creates disease in the body or it builds to a point where the person explodes with rage. So get angry, as a personal expression of anger, which does not involve other people.

Then acknowledge the buffoonery, acknowledge the extreme-duality, acknowledge the mirror of life and acknowledge the necessary-challenge. Then get over yourself (literally) and cheer up.

The easier way to cheer up is to think of something else - **distraction is the key**. Preoccupy your mind with thoughts or actions other than the previous buffoonery. When you stop re-thinking angry buffoon-type-thoughts, you will stop creating angry feelings and you will feel better.

And don't think of things that are too positive or loving or touchy-feely because the gap is too great. Think about something mundane. It will not only take your mind off the buffoonery, but mundane thoughts are slightly more positive than anger and will make you feel better. Then continue on from there with baby steps.

Cheer up as much as you can in small steps. The emotional staircase usually starts with depression and dis-empowerment and then heads upwards into higher emotions such as sadness, bitterness, anger, frustration, guilt, stress, indifference, normal, relief, hope, happiness, excitement, cheeriness, happiness, joy and finally love. Climb up the ladder, one thought at a time, one feel-better feeling at a time. The golden rule is: If your thoughts make *you* feel better than your *current* situation, then they are valid thoughts. If your thoughts make you feel nothing or worse then seek out higher thoughts.

Physical action is also a great way to cheer up, especially if you re-direct your anger or frustration into physical activity. Get busy. Attack the pile of ironing, mow the lawn or go for a brisk walk. Do something you've been putting off, ring an old friend or pat your dog (nothing beats pet therapy when you're angry). You'll feel better, bit by bit.

Once you have cheered up, keep going. Think of something funny. This will make you happy. Then think of something such as your pet, or the last great holiday you had or your new car. This will make you feel excited. Then think about your passions and inspirations. This will make you feel joy. Then think of your partner or children and you'll feel joy or love. This will reacquaint you with the feeling of joy and the practice of OPTA.

Although be careful with the partner-kids-type-thoughts. If you're not in a good enough mood, contemplating such things might quickly remind you of the day your children stuck toast in the DVD player or how your partner once made you so angry that you called them something that rhymed with a 'cupid parking patriot!'. You may quickly end up back in the bad mood again.

Once you have cheered up, consider what it was that made you angry. For example, if a person said something hurtful to you then they were being inconsiderate, rude or ignorant. This is what you *don't* want in your life, which is another reason why that event was pre-sent to you by your Soul - to teach you what you like and what you don't like in order to *know thyself.*

Now focus on the opposite - consideration, kindness and compassion. This is what you like. Now imagine a *you*niverse where everyone is considerate, kind and compassionate. These thoughts will create a feeling that is the exact opposite of the anger you previously felt.

Not only will this make you feel better but it will help you practise the thoughts required, to reverse the bad habits of thoughts within the EGO that made you react badly and feel the anger in the first place.

The added bonus is those thoughts will instantly inspire your Soul-Group to begin pre-sending events into your *you*niverse where people are considerate, kind and compassionate. If you keep thinking this way, those events will manifest into your *you*niverse as considerate, kind and compassionate, yoooge experiences.

THE ADDED BONUS OF OPTA

The practice of OPTA is so simple and so powerful … but I never said it was easy. The practice of OPTA is a real challenge that takes lifetimes to master.

People get all excited about OPTA. They try it, full-on, for a few weeks but eventually forget to think positively and soon revert back to their old ways. The reason they forget is because their decades-old EGO is far more powerful than their new, few-week-old, habit of OPTA. It only takes a few weeks (usually within the 28 day (4 x 7) natural cycle of the mind) for that bad habit to take over again. It's the subconsciousness of that habit that makes them 'forget'. It's not so much that they forgot but more that their subconscious habit is automatic and they're consciously unaware that it took over.

I've done it. You've done it. We've all done it. It's why the diet failed. It's why the New Year's resolution failed. It's why you never finished that task, rang that old friend or fully honoured that 'I'm going to be a better person' spiel you gave yourself the last time you behaved like a poorly behaved child.

Most people have *moments* of OPTA rather than OPTA. Most people can think positively for a few minutes or maybe even a few hours *at best* but the EGO eventually finds a gap and takes over to make you think negatively again and feel bad. That's its job. That feel-bad-ness squashes any chance of OPTA and creates a new necessary-challenge, where you have to cheer up again, back to normal.

Just like any other habit, you have to *persevere* to make positive-thought into a habit, so that OPTA becomes a natural part of your automatic personality; and overrides the EGO. That's a challenge! OPTA won't become a habit unless you practise, practise, practise. The more moments of OPTA you practise the more OPTA you become.

You might be thinking, 'Screw that, it's all too hard' or 'It's all a bit too touchy-feely-hippy for me' or 'I don't have time for that!' but that's just your EGO challenging you. Are you up for the challenge?

But consider this. OPTA is not just about thinking positively to feel good. OPTA is far more powerful than that. OPTA is about resolving the EGO and co-creating a better *you*niverse - a life of your dreams. Your Soul-Group loves OPTA and will work around the clock to surround your *you*niverse in OPTAmised events. Your constant practice of OPTA will create an energetic environment that will allow those events into your *you*niverse. The magnetic-potential of OPTA will attract them to you and the electrical-potential of OPTA will switch them on and they will manifest into your *you*niverse as *real events* - yoooge experiences.

You will feel happiness, joy or love, most of the time. Mind-blowing opportunity and events will come from nowhere. People around you will call you 'lucky' or wonder, 'How the %$#@ did you do/be/have that?' Physical disease will disappear. You will feel peace and calm - maybe for the first time. You'll get downloads. You'll get less necessary-challenges and be inspired towards more purposeful-challenges.

All of that can happen, simply by practising the thought-sets on page 241, as often as you can (but only when you're in good mood), so they become a natural part of your daily thinking.

SUMMARY

- OPTA stands for Only, Positive, Thought, Action.
- 'Action' gets a mention because you have to get off your arse and take positive action.
- 'Only' refers to the end goal where positive thought is the only type of thought. That will take many lifetimes.
- You cannot practise OPTA in a bad mood, nor can you use OPTA to cheer you up.
- OPTA is designed to amplify a good mood into happiness and then into joy, to re-acquaint a person with the feeling of joy and help develop a habit of consciously choosing positive thought.
- Cheering up from a bad mood *is* the challenge.

- To cheer up, acknowledge the bad mood, express the bad feelings, acknowledge the necessary-challenge, think of something else, cheer up in stages.
- Then keep going. Amplify the mood into happiness, joy and love.
- After you have cheered up, consider what made you feel bad. This is what you don't want.
- Think of the opposite. This is what you want.
- Contemplate what you want to help reverse the bad habit of thought within the EGO that made you feel bad in the first place.
- Use OPTA to co-create what you want into your *you*niverse.
- OPTA is not easy. OPTA is a challenge. It takes practice and many lifetimes to master.
- A decades-old, EGO, is far more powerful than a new, few-week-old, habit of OPTA. The EGO will take over if you don't make an effort to practise.
- Practise, practise, practise, practise, practise to make OPTA a habit.
- OPTA will co-create the life of your dreams.

THE EMOTIONAL PURPOSE

OK, so we have the four OPTA thought-sets of security, capability, expression and knowing. They are the perfect example of how to think with positive thought and they make us feel good. But they also have a specific purpose - to facilitate a purposeful-challenge. Before we go down that road, let's unravel the mindset.

CONSCIOUS/SUBCONSCIOUS INTERACTION

The four thought-sets are processed by the mind of power and the mind of wisdom. The power-mind and the wisdom-mind are often referred to as the conscious mind and the subconscious mind, respectively. You then have the conscious mind of power and the subconscious mind of wisdom. The potential-aspects of security and capability are processed by the conscious mind of power and the potential-aspects of expression and knowing are processed by the subconscious mind of wisdom.

CONSCIOUS MIND OF POWER — Security, Capability

SUBCONSCIOUS MIND OF WISDOM — Expression, Knowing

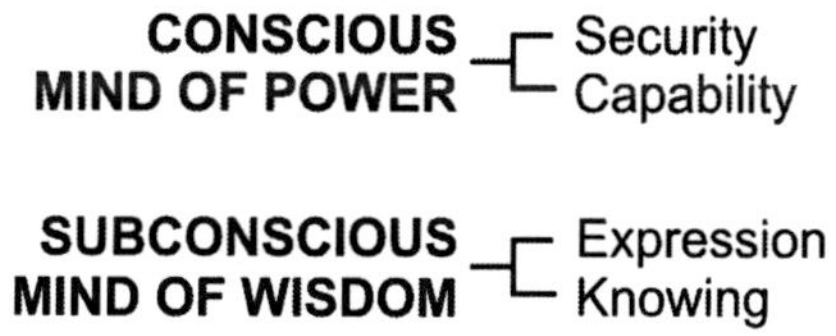

The conscious mind is a mind of now - what's happening right now in life and what do we do about it. That part of life is best handled with security and capability. That's why the potential-aspects of security and capability are contained within the conscious mind and influence the brain to consciously think security and capability thoughts, *now*. These are physical-world, action-based, types of thought. The physical-world represents reality, which comes to us moment by moment, as moments of *now*. Consequently the power-mind becomes the mind of now, otherwise known as the conscious mind of power, or the *power of now*.

The subconscious mind is our automatic mind. It operates without our conscious consideration and takes care of all the 'repetitive' thought-actions, such as instinct, talent, speech, personality and imagination. But to get its automation, it has to assemble a lifetime of thought-actions and then categorise the thoughts behind those thought-actions into habits of thought-action. That's why the role of the subconscious mind is to record every thought we have ever made. Those thoughts accumulate in the subconscious mind to become automatic habits of thought - memory. The potentials of expression and knowing stored in the subconscious mind then help you to retrieve those memories as automatic habits of thought. Those habits help you wisely *react* to life (*now)*, based on the memory of how you have previously *responded* to life (*not-now)*.

For example, secure thought-action *now* is expressed as survival and adventure. The *memory* of secure thought-action is what makes you trusting and courageous. Capable thought-action *now* is expressed as skill and willpower. The *memory* of capable thought-action is what makes you talented and successful. The *memory* of our previous security and capability gets added to our memory of expression and is then re-expressed automatically (reaction) as our personality, communication and relationships. The *memory* of our security, capability and expression is then added to our memory of knowing which feeds our life-experience, creative-intellect and imagination.

Characteristics such as trust, courage, talent, personality or imagination are automatic. We don't consciously think about them because we use the subconscious mind to feed our personality or imagination. Consequently the wisdom-mind becomes the mind of not-*now,* otherwise known as the subconscious mind of wisdom or the *wisdom of not-now* (memory).

The conscious mind of power and the subconscious mind of wisdom, work together to enhance the overall function of the mindset. For example, you might notice a car broken down on the side of the road with a flat tyre. You then consciously respond to life by thinking, 'I am going to stop and help them change that tyre'. That decision is a conscious decision made by you. It's a secure (brave and trusting) and capable (skilful, confident and willing) thing to do. It takes a good amount of personal power to make such a decision. It's a powerful response made by your conscious mind of power.

That decision is based on decades of memories stored in your subconscious mind that now forms your expression and knowing. You have the trust and confidence to easily **express** yourself from decades of security and capability. This makes you a friendly helpful person who is communicative enough to approach strangers and help them. It's who you are and you can easily express this. You also have the **knowing** memory from helping people in the past that knows you are **secure** enough to trust people, **capable** enough to help people, you can **express** that trust by being friendly and your **knowing** that you love helping people. It's also a very wise decision you've made. It's a wise reaction made by your subconscious mind of wisdom.

So you pull over and decide to help this person. As you do, your conscious mind of power is constantly monitoring the situation to keep you safe (security) and focussing on the task at hand with skills required to change a tyre (capability). Your subconscious mind of wisdom is constantly retrieving the automatic habits of thought required to communicate to a stranger (expression), change a tyre (knowing-skill) and react to danger (knowing-life-experience).

Good responses to life (and reactions since OPTA is also a habit of thought) with powerful and wise thought form the basis of OPTA. Bad reactions to life with negative thought, forms the basis of EGO (unless you're Dr Evil, it's unlikely you'll consciously respond - choose - negative thoughts). Good responses accumulate into a habit of OPTA. Bad reactions accumulate into a habit of EGO.

The conscious mind becomes the 'I am ...' part of the 'I am that I am' mindset and the subconscious mind becomes the '... that I am' part of the 'I am that I am' mindset. The pulse of the mindset combines the twin mind functions of the conscious mind and the subconscious mind back together to form 'I am ... that I am'. Your human-mindset then becomes a living breathing expression of original 'I am that I am' thought of GOD. Ponder that!

- I am conscious that I am subconscious.
- I am now that I am not-now.
- I am power that I am wisdom.
- I am security that I am capability.
- I am expression that I am knowing.
- I am EGO that I am OPTA.

The conscious mind of power exists within the physical-world and the subconscious mind of wisdom exists within the astral-world. We can't see or touch the conscious mind because it exists at a frequency beyond our perception (similar frequency to the human aura - it's one and the same thing) but its energetic platform operates in the physical-world.

That's why you're more familiar with your conscious mind. It's physical, it can be controlled by you and its presence can be experienced via the brain. It's the mind you associate with your own self, or the *you that you **think** is you* (*you* for short). The subconscious mind is more of a mystery. It's non-physical (astral), cannot be controlled by you and exists at a frequency beyond human perception. It's a hidden, automatic, mind function.

E-MOTIONAL PURPOSE

I know that some readers, by now, have probably had enough of all this pulsing, sciency, mind stuff. Some of you may be confused, dizzy or even losing the will to live. But you've gotten this far, so hang in there. If you're feeling sleepy, pull out a nose hair. It brings your focus back to the moment of *now,* immediately. It may fill your eyes with tears, so not a good idea whilst flying a plane. The next few chapters help tie everything together. This is the fun bit. Then we can move onto ghosts, 2012 and death. Yah! In saying that, here's some more pulsing, sciency, mind stuff.

When we think, the human-mindset pulses back-and-forth between the physical-world and the astral-world, nonillions of times a second. The rapid nonillionth-of-a-second pulse between the physical-world and the astral-world, re-combines the *electrical*-potential behind the security-thoughts with the *magnetic*-potential behind the capability-thoughts, to create *electro-magnetic* thought-waves of power. The pulse also combines the *electrical*-potential behind any expression-thoughts, with the *magnetic*-potential behind any knowing-thoughts, to create *electro-magnetic* thought-waves of wisdom.

These electromagnetic thought-waves create ripples in the Cosmic-pond (ripples in a pond). As mentioned before, thought-waves don't travel. They bypass space-time and are everywhere-all-at-once. When they collide with the thought-waves of other people, they become points of vibrational-reality that co-create your *you*niverse. But they also do something else quite extraordinary. The ripples from your thought-waves flood your body with electromagnetic vibration and that creates ***emotion***!

The ripples in the Cosmic-pond created by your thought-waves are potential energy-in-motion or e-motion for short. E-motion floods the body with electromagnetic ripples, everywhere-all-at-once, which we experience as the real, physical sensation of *feeling*. That's why feelings tingle.

We feel the buzz of our thoughts! You can feel it in your body. You can feel it in your blood. You can feel it in your nerves.

It's important to make the distinction between e-motion and feeling. E-motion is energy-in-motion, which are the thought-waves. Feelings are the physical sensations created by those thought-waves as they flood the body. E-motions are the waves, feelings are the result.

Pulse - thought - thought-waves (e-motion) - feeling

Thought-waves of power create the e-motion of *desire,* which we experience as *feelings* of courage, abundance, passion, willpower and success. This becomes the lower emotional centre. Thought-waves of wisdom create the e-motion of inspiration, which we experience as *feelings* of individuality, fulfilment, intuition, purpose and knowing. This becomes the upper emotional centre.

Below is a more complete list of the more common feelings associated with the e-motions of desire and inspiration:

E-MOTION OF DESIRE	E-MOTION OF INSPIRATION
Creates feelings of:	**Creates feelings of:**
Safety	Significance
Courage	Individuality
Abundance	Fulfilment
Passion	Intuition
Willpower	Purpose
Achievement	Creative-intelligence
Self-worth	Knowing

(Table) The more common feelings associated with the e-motions of desire and inspiration.

Our feelings combine with our thoughts to produce a sort of 'oomph' motivational force. For example, the thoughts of security and capability (power) combined with the feelings of courage, abundance, passion, willpower and success, create a *powerful-desire* to do stuff. The thoughts of expression and knowing (wisdom) combined with the feelings of individuality, fulfilment, intuition, purpose and knowing, create an *inspirational-wisdom* to think about stuff (it's actually supposed to read 'wise-inspiration' but 'inspirational-wisdom' sounds better).

These two 'oomph' forces combine to give us *purpose* - a desire to do stuff combined with the inspiration to think about stuff. It's that purpose that drives us towards a purposeful-challenge.

When you think about it, what makes you want to be a better person, succeed in your career or participate in a fun hobby or sport? It's your innate powerful-desire to do stuff combined with your inspirational-wisdom to think about stuff. That comes from your thought-action of security, capability, expression and knowing. That becomes your default power and wisdom. That's why you have a conscious mind of power and a subconscious mind of wisdom. That's why you're here as a human *being* GODlike.

THE MOST ENJOYABLE WAY

The most enjoyable way to satisfy your powerful-desire to do stuff combined with your inspirational-wisdom to think about stuff is **doing what you love because you know what you love doing**, whether that's your career, hobby, vision, day-job or calling. Doing what you love is the highest form of 'a powerful-desire to do stuff'. Knowing what you love doing is the highest form of an 'inspirational-wisdom to think about stuff'. In short following your dreams. The only reason your thoughts co-create events in your *you*niverse is so you have everything you need to follow your dreams.

But you can't *do* what you love until you *know* what you love doing. You can't *know* what you love until you *do* what you love. That's where your powerful-desire and wise-inspiration come in. Powerful-desire makes you do stuff so you know what you love doing. Inspirational-wisdom makes you think about stuff so you do what you love.

Power comes from thoughts of security and capability. Security gives you the courage to explore the world, go on adventures and try new things. It also helps you to trust people and the abundance of security helps co-create everything you need to make that happen. Capability gives you the energy and willpower to do stuff. It helps you learn new skills and achieve your goals. It helps you make the right choices, including choosing the desirable course of action and the right people to help you. It also gives you the confidence and self-worth to accept success. **It's how you do what you love.**

As soon as you do stuff, you immediately experience, first hand, what you like and what you don't like. And in a world of extreme-duality, there is then plenty of choice. Those experiences accumulate in the subconscious mind to add to your expression and knowing.

Expression helps you personally and creatively communicate your likes and dislikes to the world, so that other people know who you are and what you want. People can't help you until you express what you want/like/need. Expression creates change, makes things happen and adds a creative voice to your inspirational-wisdom. Expressing your ideas to others helps maintain your importance and validate your ideas or vision. Knowing includes your life-experience, imagination and creative-intelligence (it's how you know). Your life-experience helps you re-member what you like and don't like. Your imagination inspires you to do more of what you love and to try new things. It fuels your desires and keeps you inspired. Your creative-intelligence helps conquer the challenges of doing stuff, solve problems and be creatively unique. Knowing also helps you express your desires with compassion and gratitude, in order to serve others. **It's how you know what you love doing.**

The powerful-desire of the conscious mind combined with the inspirational-wisdom of the subconscious mind urges you to do what you love because you know what you love doing. It's life-force and it literally keeps us alive.

We've all felt the desire and the inspiration of this life-force. Our powerful-desire drives us and our inspirational-wisdom inspires us. It's what gets us out of bed in the morning. It's what compels us to be a better person. It's what fills our eyes with tears of joy. It's why we all worship potentially famous people. It's what makes us feel miserable, regretful or guilty when we ignore it.

Using your powerful-desire and inspirational-wisdom to do what you love because you know what you love doing is also the greatest purposeful-challenge. It's the most rewarding and challenging way to re-member your forgotten potential. It's the most enjoyable way to practise OPTA. It's the fastest way to resolve your EGO. It's the path to en-Light-enment. Live your dreams.

SUMMARY

- The power-mind is the conscious mind.
- The wisdom-mind is the subconscious mind.
- The power-mind contains the potential-aspects of security and capability.
- The wisdom-mind contains the potential-aspects of expression and knowing.
- The conscious mind and subconscious mind work together to enhance the overall function of the mindset.
- You consciously respond to life now, with powerful thoughts of security and capability.
- Those responses are stored in the subconscious mind.
- You then react to life with expression and knowing, based on the memory of how you have previously responded to life - habits of thought.
- Good responses accumulate into a habit of OPTA.

- Bad reactions accumulate onto a habit of EGO.
- The conscious mind of power exists within the physical-world.
- The subconscious mind of wisdom exists within the astral-world.
- When we think, we pulse back-and-forth between the physical-world and the astral-world.
- Thoughts create ripples in the Cosmic-pond that are everywhere, all at once.
- We experience these ripples as feelings.
- Thought-waves of power create the e-motion of desire, which we experience as feelings of courage, abundance, passion, willpower and success.
- Thought-waves of wisdom create the e-motion of inspiration, which we experience as feelings of individuality, fulfilment, intuition, purpose and knowing.
- Thoughts of security and capability (power) combined with the feelings of courage, abundance, passion, willpower and success, create a powerful-desire to do stuff.
- Thoughts of expression and knowing (wisdom) combined with the feelings of individuality, fulfilment, intuition, purpose and knowing, create an inspirational-wisdom to think about stuff.
- The powerful-desire to do stuff combined with the inspirational-wisdom to think stuff is what drives us towards a purposeful-challenge.
- The most enjoyable way to satisfy your powerful-desire to do stuff combined with your inspirational-wisdom to think about stuff is **doing what you love because you**

know what you love doing, whether that's your career, hobby, vision, day-job or calling.

- You can't *do* what you love until you *know* what you love doing.
- You can't *know* what you love until you *do* what you love.
- That's why you have a powerful-desire to do stuff and the inspirational-wisdom to think about stuff.
- Powerful-desire makes you do stuff so you get to know what you love doing.
- Inspirational-wisdom makes you think about stuff so you do what you love.
- That's why you have a conscious mind of power and a subconscious mind of wisdom.
- Doing what you love because you know what you love doing is the greatest purposeful-challenge.
- It's the most rewarding and challenging way to re-member your forgotten potential.
- It's the most enjoyable way to practise OPTA.
- It's the fastest way to resolve your EGO.
- It's the path to en-Light-enment.
- Live your dreams.

JOY

When you embrace your powerful and wise potential, your conscious mind of power will respond to life with thoughts of security and capability, which will create the emotion of desire. The e-motion of desire will flood your body with various mixed feelings, such as courage, abundance, passion, willpower and success. The subconscious mind of wisdom will react to life with thoughts of expression and knowing, which will create the e-motion of inspiration. The e-motion of inspiration will flood the body with various mixed feelings, such as individuality, fulfilment, intuition, purpose and knowing.

These thoughts and feelings together urge you to do what you love because you know what you love doing, as a purposeful-challenge. The human-level pulse between the physical-world and the astral-world will then combine all these feelings together so that you will *simultaneously feel* courage, abundance, passion, willpower, success, individuality, fulfilment, intuition, purpose and knowing *all at the same time.* You will feel a gigantic super-feeling otherwise known as *joy.*

THE FEELING OF JOY

The feeling of joy is a big feeling. To help illustrate just how big the feeling is, the following table shows some of the more common feelings that make up the super-feeling of joy. It's different for everyone and the feeling combinations change, so the following table is more of a guide. Generally speaking, to feel true-complete-joy, you would have to *simultaneously* feel most, if not *all*, of the following feelings, in equal and balanced proportions.

You would have to feel:

E-MOTION OF DESIRE	
Security	**Capability**
Courageous	Energetic
Trusting	Empowered
Supported	Guided
Excited	Appreciated
Safe	Successful
Adventurous	Worthy
Abundant	Skilled
Supplied	Accomplished
E-MOTION OF INSPIRATION	
Expression	**Knowing**
Important	Purposeful
Fulfilled	Imaginative
Understood	Compassionate
Communicative	Intuitive
Individual	Peaceful
Validated	Spiritual
In control	Grateful
Principled	Creatively-intelligent

(Table) The Feeling Table of Joy. A person would have to feel most, if not all, the feelings above to experience the feeling of joy.

That's joy - true and complete joy. That's a big feeling. It can be so overwhelming it can fill your eyes with tears. It can make you grin incessantly to the point that people will think you're weird or drugged. It can create a state of excitement and expectation that is almost impossible to bear. It can fill you with such a profound level of gratitude that if your eyes are already filled with tears of joy then you'll start blubbering and laugh/crying with joy. But that's true and complete joy. There are not many humans on the planet that can feel this level of joy.

WHAT JOY IS NOT

The true and complete feeling of joy is beyond most people. Most people rarely feel joy, if at all. Joy is a state the human race is heading *towards* but currently does not feel *en masse*.

To better explain what joy is, it's easier to explain what joy is not. When people read about joy they often say, 'Oh yeah, I'm a joyful person!' But you can't be a joyful person unless you are *simultaneously* feeling most, if not all, of the feelings described in the previous 'Feeling Table of Joy'.

To simultaneously feel most of the feelings described in the 'Feeling Table of Joy' you have a perfectly balanced mindset of power *and* wisdom, containing equal levels of all four potential-aspect aspects (that will depend on your memory-of-potential and what potential you have remembered in this lifetime). You must then equally embrace all four potential-aspects to think with the security, capability, expression and knowing required to create identical amounts of desire and inspiration. You must then have a body healthy enough to handle the electromagnetic vibration of all the aforementioned feelings. Only then will you experience the super-feeling of joy.

That's what the word 'enjoy' actually means. The prefix 'en' means 'to encircle' and 'joy' comes from power *and* wisdom. So to en-joy yourself means to embrace your powerful *and* wise potential and *encircle* yourself in powerful and wise thought to create the e-motions of desire *and* inspiration required to feel joy. That's why I often write the word 'enjoy' as 'en-joy'.

Unfortunately, most people don't have that balanced mindset. Few humans do. The EGO produces all sorts of imbalance in the mindset. That imbalance deranges the potential conditions needed to experience joy. Imbalance is always unique to the individual. Never are two people ever screwed up the same way! That ensures extreme-duality of impurity. But there are certain trends.

For example, there is often imbalance within the potential-aspects of each mind where one potential-aspect is at a higher level than the other. This creates unbalanced thought-action-emotion-feeling. For example, being capable without security creates apprehension or self-doubt. Being secure without capability creates failure. Being expressive without knowing creates ignorance. Being knowing without expression creates withdrawal.

There is often imbalance within the mindset as well, which again ultimately leads to an imbalance in thought-action-e-motion-feeling. In this example if we focus on purposeful-challenges then people with too much wisdom and not enough personal power tend to be inspirational but without desire. This often means these people have great ideas but are too scared to take the plunge, or can't be bothered, or lack the necessary skills to make it happen. Consequently the purposeful-challenge never gets started or is rarely completed. That usually creates boredom and depression.

People with too much personal power and not enough wisdom tend to be full of desire but without inspiration. This often means these people will leap into a purposeful-challenge but without proper planning, experience, vision or the personality required to pull it off. Consequently the purposeful-challenge often fails or becomes mismanaged. This creates frustration and failure.

In each and every case of such imbalance, people will not feel joy. They may experience *moments* of joy, from highly charged events in their lives that inspire enough potentiality, emotion and feeling to create a moment of joy but rarely a fully joyous experience. Moments in life that fall into this category include the birth of a child, a successful completion of a lifetime goal, falling in love or a peak spiritual experience.

GENERAL-UNHAPPINESS

The imbalance in people's mindset usually always leads to part-joy. Part-joy is a feel-good feeling that's produced from an odd assortment of partially embraced potential-aspects, which then produces a diluted version of desire or inspiration; which then creates an incomplete combination of feel-good feelings that we experience as happiness.

Consequently, there are many versions of happiness because there is no end to the potential-emotion-feeling combinations of an imbalanced mindset. Happiness includes all the feelings that make you smile, laugh, go 'Orrr' as well as contentment and peace. But it's not joy. Not even close! Joy includes all the feelings that make you grin like a silly person, belly laugh, go 'Woohoo' and cry. Joy is a feeling way beyond any 'normal' human expectation of happiness.

And believe it or not, the *rest* of the western-world isn't even *that* happy and most westerners exist in a slightly negative state of general-unhappiness.

I keep referring to westerners and the western-world because I am generalising and do not want to include non-westerners in the statement 'most people do not feel joy'. It offends them. There is so much more en-Light-enment in the Eastern countries of the world and some of the most joyful and potentially en-Light-ened people live in abject poverty in Third World countries - believe it or not. The western-world I refer to includes countries such as Australia, the United Kingdom, The United States, Canada, Europe or any other country whose residents worship pop stars, drink copious amounts of beer, kill animals for sport or spend billions of dollars on Botox™ injections.

General-unhappiness starts with life-laziness and most of the western-world is life-lazy. Everything about the western world has been designed to reduce the challenge, which makes us life-lazy. We spend billions of dollars on gadgets that remove the challenges once faced by our ancestors. We harnessed electricity because we disliked the challenge of lighting a fire for heat and light. We developed the car because we disliked the challenge of maintaining a herd of horses. We developed modern medicine because we disliked the challenge of natural healing and remedies.

Don't get me wrong. I love the technology of modern life as much as the next person. I'm ever so grateful that my iPod™ doesn't involve charcoal, horse poo or steeping herbs in a hot broth. It's just that there is not enough potential en-Light-enment in our modern lives and so the western-human-race, especially in the last one hundred years, has become life-lazy.

Because we don't challenge ourselves enough the western-human-race has become overpowered with EGO. It's what's fed the wars, poverty and crime of the last century. It's the reason why disease is at epidemic proportions (desire and inspiration *is* life-force). It's why we're glued to our TV sets worshipping rock stars, film stars, actors, sports people and reality television shows. We live vicariously through them via the TV screen because we secretly want to be them because we subconsciously yearn for the potential within them. If we all challenged ourselves at that level, then those sorts of people would seem 'normal' to us and we wouldn't be so addicted to their stories.

It doesn't mean we have to climb Mt Everest, own our own airline or write a hit song. Being a good parent is more challenging than all of those things. It's about purposefully challenging ourselves, *just for the sake* of doing what we love because we know what we love doing.

But most people don't do that. Most people don't 'step out of their comfort zone'. Most people do what's easiest. They do what scares them less. They do what makes them the most money. They do the minimal amount to get by. They do what others or society expects them to do. They do what gets them the most approval. They do what their father or mother did. They do what their culture or religion tells them to do. They do what they do to avoid conflict or attention. They do what they do because they don't know what they love. They do what they *have to do* when faced with a necessary challenge. In short, they do what their EGO tells them to do and it represents ninety percent of the western-world.

Rarely do people do what they love because they know what they love doing and *only* because the purposeful-challenge gives them joy. I'm sorry, but it's true, whether that's personally, professionally, recreationally, as a parent, or with friends and family.

Meanwhile the other ninety-five percent will argue until spittle runs down their chin that they *do* what they love, they know what they love and that *they are* happy. They may do what they love but it's more likely they do it because it's what their EGO thinks they should do and fulfilling the EGO produces a temporary happiness. They say they know what they love but their EGO has most probably limited them to knowing only what the EGO is willing to let them try. They may be happy but they're not grinning like silly people as they belly laugh, go 'Woohoo' and then burst into tears of joy and so it's not true-complete-joy.

If it's not joy then it can't be a balanced mix of potential. If it's not a balanced mix of potential then they can't be *fully* doing what they love because they *really* know what they love doing. They can have rare moments of joy and lots of moments of happiness and even consider themselves to be happy-go-lucky people but their EGO has them trapped in a negativity feedback loop.

They lack potential, the EGO takes over, they feel bad, they go downhill into a feel-bad pit of negativity (voluntary journey into impurity), they eventually cheer up back to 'normal' (voluntary journey back to purity), only to have their EGO react badly and send them downhill again into the same feel-bad pit of negativity.

They go up-down, up-down, up-down on an emotional roller-coaster, repeating the same mistakes over and over again, in a general state of unhappiness. There is growth within that cycle, from the natural increase of potential within every spiral, but it's *slow* like a 'snail' and involves unnecessary pain and suffering.

They think they're happy but *really* their emotional roller-coaster is mostly bouncing off the *lower limits* of happiness keeping them trapped in a general state of unhappiness. They've become so used to this emotional state that they consider it to be 'normal' and so assume they're happy. But it's not 'normal' and they're not happy. It's general-unhappiness created by the EGO taking over, due to a lack of potential challenge.

They lack potential. They've become life-lazy. They're trapped in a negativity feedback loop. Without the potential gained from purposeful-challenges they won't be able to think the powerful and wise thoughts required to bust out of this negativity loop. Their Soul will be 'forced' to pre-send them more and more necessary-challenges, at a level that becomes more and more drastic. Those necessary-challenges will make them react to life even worse, they'll go downhill even further until they can't stand it anymore and decide to *do something about how they feel.* Only then will they change their thoughts, feel better, wake up and bust out of their negativity feedback-loop.

All this could be avoided if they only decided to do what they love because they know what they love doing.

Do you know why the majority of people who watched the movie ‘The Secret’ and then tried to manifest a million dollars, didn’t even come close? Apart from their bad habits of negativity preventing their dreams from manifesting into their *you*niverse, most people wanted that money to buy more life-laziness. They wanted to work less, do less and strive less. That’s what money can buy you but there is *less* challenge in that.

It’s no surprise that between eighty and ninety percent of instant lottery millionaires in the United Sates end up broke and miserable within a year and would do anything to have their old lives back. That’s because their Souls present them those broke and miserable necessary-challenges to counteract the life-laziness that their lottery money bought.

If these same people had instead focussed on maintaining a habit of positive thought and then used that money to fund an en-joyable and challenging purpose, then their lives would have been completely different. However, if the same people already *had* such a life then that challenging lottery win would never have been pre-sented to them in the first place.

How many people do you know who truly *love* what they do for a living, to the point that they would work in their day job without being paid. How much do you love *your* job? Do you jump out of bed, eager to get to work? Are you the best person you can be? Are you a really really good parent/partner/friend? Do you judge others, criticise yourself or feel guilty? How much free time do you devote to an all-consuming passion or hobby? Have you ever followed your *dreams?* When was the last time you trusted in yourself enough to take a chance? What do you want to do, have or be? Who will you be, where will you be and why will you be there in ten years? Twenty years? Forty years?

THE PURPOSE OF JOY

The purpose of life is to do what you love because you know what you love doing - because it *feels good*. Your desire and inspiration gets you there. Your powerful-desire drives you and your inspirational-wisdom inspires. Joy is the reward. Joy feels amazing and there is not one single human-being on the planet that does not want to feel good. The only reason we think/do anything is because it feels good and the only reason we avoid things is because it feels bad.

There is nothing more important than how *you* feel, hence why the purpose of life is to *feel good* by doing what you love because you know what you love doing. Life is all about you, *first*. Joy is all about *you*. It's not selfish it's being true to self. Following your dreams is all about *you*. The next step up from joy is love but that comes later. Joy is the most important step for *you, now*.

The more you do what you love because you know what you love doing, the more you will do, the more you will know, the more moments of OPTA you will have to practise, the more en-Light-ened you will become, the more joy you will feel.

That's why the Gods instilled an emotional compass right in the middle of your chest, otherwise known as feeling. Life is a challenging journey back to en-Light-enment. It can be confusing and you can get easily lost on your path. You may not be aware of your thoughts (ninety per cent of thought can be subconscious). You may not be aware of your actions (body language, posture, weird habits, etc) but you CANNOT mistake how you feel. That's why your emotional centre is placed in the centre of your chest.

Feelings are always there. You CANNOT block feelings anymore than you can block soundwaves coming out of a speaker. Feeling is the physical experience of vibration. You can ignore them but you cannot block them. You can also feel feelings in your gut area or head area but the true feeling is heart-felt in the centre of your chest.

If you pay close attention to how you feel (not what you *think* but how you feel) then you will always know whether what you are thinking is powerful and wise, or weak and foolish. Powerful and wise thought makes you feel good, great, fantastic, awesome or tearfully joyful, (depending on how much potential you embrace), so head in that direction. Weak and foolish thoughts will make you feel bad, serious, awful, terrible, miserable or tearfully suicidal (depending on how much you resist your potential and what sort of EGO you have), so don't head in that direction.

Life is really quite simple! Follow your feelings. Feel your way. For example, powerful and wise thoughts of security, capability, expression and knowing all feel great. Purposeful-challenges feel great. Doing what you love because you know what you love doing feels even better. Happiness is wonderful. Desire and inspiration feels awesome. OPTA feels amazing. Joy will fill your eyes with tears. Love is unmistakeably GODlike.

Foolish and weak thoughts of insecurity, failure, depression and ignorance all feel bad. Necessary challenges often feel bad. Not doing what you love because you don't know what you love doing is boring, soulless and frustrating - feels bad. Apathy and ignorance feel worse. The EGO can make you feel 'just not quite right', too shallow, unfulfilled, bad, awful or downright miserable. Seriousness (the opposite of joy) is unmistakeably unGodlike. Hate feels like crap!

And a 'heads up' about the EGO. It's sneaky. The EGO is an integral part of your subconscious personality and spiritual journey. It's not to be controlled, judged or blocked. It's part of you. It is you, so embrace the EGO for what it is - an impurity creation mechanism helping you find en-Light-enment. The EGO utilises outdated childish habits of weak and foolish thought that include your subconscious fears, failings, depressions and ignorance. Those types of thoughts often make you feel bad and go downhill. But the EGO can often make you feel good … in a sneaky kind of way.

The EGO is like an old friend, who has been with you forever but is a bit of an arsehole. The old friend sometimes treats you badly, can be quite rude, sometimes offensive, mostly selfish and very unaware. But you love that friend just the same and you put up with his/her crap because you know that he/she is always there to help you when you needed it (en-Light-enment). You might ignore your own needs, you might purposefully loose to make them look better, you might avoid standing up for yourself because you can't be bothered with the conflict, or you might choose to keep quite even though you know better. And you do all of this, just because it's easier.

The EGO is very similar. When we avoid life due to fear, give up because we don't want to fail, suppress ourselves to avoid conflict or choose 'ignorance is bliss', it often makes us feel better because those thought-actions satisfy the bad, subconscious habits of thought that make up the EGO. It's almost like we're doing what the EGO *thinks* is the right thing to do and this can generate a feeling of safety, comfort, peace or humility. This is what life-laziness feels like and why life-laziness can be so addictive. It's the 'comfort zone'.

But it's all false. These feelings aren't heart-felt, they lack substance, they're always temporary and so fade and your intuition (deep down) knows they're not real. It just doesn't feel right. It's all EGO and the fastest way out of the EGO is to step out of this 'comfort zone' with the purposeful-challenge of doing what you love because you know what you love doing, and *only* if it gives you joy. Feeling good is not enough because feel-good can be EGO. Joy *transcends* the EGO, which is why it's so important to practice OPTA.

THE EMOTIONAL COMPASS

Your feelings work like an emotional compass. If you feel good, you know you are heading in the right direction towards purity of thought and en-Light-enment. If you feel bad (or not quite right), you know you are heading in the wrong direction towards impurity of thought. The EGO can sometimes make that compass hard to read but if you just clean the glass by paying more attention to *how you feel* (including your subtle intuition), you can better distinguish between true-feelings and the fake-feel-good feelings of the EGO.

Even though you have an emotional-compass in the centre of your chest, you are free to ignore it. You can ignore your true-feelings (many do) and go in any direction you choose. That's why you have free will. You can change your thoughts at any time (good or bad) and change your direction at any time. The Cosmos doesn't care which way you go because good or bad, it's all needed for the sake of extreme-duality.

But consider this. Reacting to life badly with EGO is the wrong direction to travel. In that regard it's *backwards.* It's living backwards rather than moving forwards. That's why it feels so bad. What's really interesting about living your life backwards is if you spell the word 'live' backwards you get the word 'evil'. Similarly, if you spell the word 'lived' backwards you get the word 'devil'. Feeling bad is the true meaning of evil. In doing so you become the devil and create 'Hell on Earth'. Joy is 'Heaven on Earth'.

THE GAME OF LIFE

Joy is a state that has to be strived for. It has to be earned. It takes years of personal growth and *lifetimes* to master - especially considering that mastering joy is not about being joyful but being joyful *all the time.* That's why I keep mentioning OPTA because a habit of consciously choosing positive thought is what's required to maintain a constant state of joy.

That real challenge of mastering joy lies not in trying to maintain a constant state of joy but trying to do so when everybody else around you is being so fricken serious, worrying themselves to death in an EGOic state of general-unhappiness. That's a tricky, fascinating, and rewarding purposeful-challenge in our life-lazy western-world. I think it's why monks live in solitude.

Mastering joy is what I refer to as the 'Game of Life'. You complete each *level* in the Game of Life when you re-member enough powerful and wise potential, to think newer more powerful and wise thoughts, powerful and wise enough to overpower the old habit of negative thought, within the EGO. Just like the layers of an onion, each new level of potential resolves a layer of the EGO.

You then move onto the next level. Layer by layer, you slowly resolve the EGO with more and more moments of OPTA, creating more and more moments of joy. These moments of joy all string together to help maintain a constant (or more constant) state of joy - within the generally-unhappy, fricken-serious world. The reward for finishing each level of the 'Game of Life' is joy. *You* feel better each layer and that's all that matters. It's all about you.

YOU USED TO KNOW

The good news is you know this state of joy. You actually used to practise this state of joy daily before it was 'knocked out of you' by a bunch of serious adults. This state of joy was called childhood. Remember when you were little? Remember the adventure of play? Remember the wonder of the worlds you created with your imagination?

Remember the excitement of new discoveries? Remember Christmas? That's joy.

My niece, when she was three years of age, would sometimes experience so much joy that her little body couldn't handle it. She'd just stand there, looking at you, unable to move, trembling, shaking her clenched fists up and down, red-faced until the joy-build-up was too much and she would explode with a squeal of joy.

You can still see this joy in most young children. If you have children, watch them play. If you don't have children then visit a child's playground and watch other people's children play. Be careful though. The days of standing in a playground, grinning as you observe other people's children, have long gone. It's more likely that Police, phoned in by the concerned parents, will arrive and make a public spectacle out of you by yelling, 'You! Yes you! Beside the 'Wiggles' swing set. Stop smiling! Put your hands in the air and step awaaay from the children!'

Watch children play. Observe the sheer joy on their faces. They don't worry about bills, relationships or work. They just play, in a *constant state of joy* - mostly. This used to be you. What happened?

The good news is that deep down, within your psyche is the memory of this joy. You don't have to learn it, you just have to re-member how it feels - hence the practice of OPTA (not to cheer you up but to amplify feel-good feelings into joy so you re-member the feeling of joy). That's why we promote the art of childish silliness, in adults. When adults act like children, and play joyfully, they forget to be adults for a moment and that childhood joy returns.

Don't believe me? Try skipping (when no else is looking). Not with a skipping rope but bounding, one leg at a time, across the floor. Children do this all the time and it's always playful and fun. You would have skipped as a child. There is an associated memory locked in your subconscious mind that skipping equals joy. It's now impossible for you, as an adult, to skip without smiling. Even more so if you hold the hand of your partner and skip. Try it!

Children are the best teachers of joy - especially the generation of babies being born now. *Your joy* has long been replaced with EGO. Babies are born without EGO. That's why they are so delicious and you want to eat them up (parents know what I mean). Babies are born pure. You can see this strange Divinity within them. Even those babies who have grown into children are still learning their bad habits and so are still more joyful than most adults. These future children will play a major role in the en-Lightenment of the planet, which starts in 2012.

For many reasons, be a child, be silly, play as much as you can, cherish Christmas and poke your tongue out at authority.

THE YOOOGE SYMBOL

The Yoooge symbol came to me in a vision (download), some years after the events of November 2003. I didn't realise it at the time (I just thought it was a pretty picture) but the Yoooge symbol actually represents all the elements required to create the feeling of joy.

The lower circle of the symbol represents the lower-aspect of power and the upper circle of the symbol represents the upper-aspect of wisdom. The circle crossing over in the centre represents the potentials of power and wisdom re-combining to create joy. The horizontal line represents the zero-point-centre of nothingness (midline) re-combining the potentials back together during the human-level-pulse. The eye represents GOD who is observing the entire process to feel love.

The lower and upper circles also represent the extreme-duality of security/capability (the eye is desire), expression/knowing (the eye is inspiration), physical-world/spirit-world (the eye is the astral-world), conscious mind/subconscious mind (the eye is the breath) and space/time (the eye is the Cosmic-pond).

(Figure) The Yoooge Symbol.

SUMMARY

- The human-level pulse between the physical-world and the astral-world combines all the feelings created by desire and inspiration *together* to create the super-feeling of joy.
- Joy is the combined feeling result if you simultaneously feel courage, abundance, passion, willpower, success,

individuality, fulfilment, intuition, purpose and knowing all at the same time.

- Most people don't know what joy is.
- Joy is all about you. To en-joy yourself means to embrace your powerful *and* wise potential to create the e-motions of desire *and* inspiration required to feel joy.
- Joy can only come from a balanced mindset.
- Most people have an imbalanced mindset because the EGO creates imbalance in the mindset.
- That imbalance creates part-joy, otherwise called happiness.
- Most people don't even feel this happiness.
- Most people exist in a state of general-unhappiness. They have become so used to this state that they believe it to be happiness.
- Technology has made westerners life-lazy. They lack challenge and consequently the EGO has taken over in the last 100 years.
- Most people don't purposefully challenge themselves or 'step out of their comfort zone'.
- Rarely do people do what they love because they know what they love doing.
- They lack potential and so have become trapped in a negativity feedback loop.
- Their Soul is then pre-sending necessary-challenges to help bust them out of this negativity feedback loop.
- The purpose of life is to do what you love because you know what you love doing, because it feels good.
- There is nothing more important than how you feel.
- That's why you have an emotional compass in the centre of your chest.

- Powerful and wise thought makes you feel great - go that way.
- Weak and foolish thought makes you feel bad - don't go that way.
- Follow your feelings.
- The EGO uses weak and foolish thought and so feels bad.
- But the EGO is sneaky and can make you feel good in a false kind of way.
- OPTA and joy transcend the EGO.
- If you pay attention to how you feel you will always know which way you are heading on the path.
- Forwards towards purity of thought and en-Light-enment or backwards towards the 'evil' of feeling bad.
- Life is a game and the real challenge is not joy but maintaining a constant state of joy.
- You master each level of the 'Game of life' when your moments of OPTA are powerful and wise enough to resolve an old habit of the EGO.
- That's why the practice of OPTA is so important.
- You used to know joy - it's called childhood - you just have to re-member it. Another reason why the practice of OPTA is so important.
- The Yoooge symbol represents all the elements required to en-joy yourself.

'Follow your bliss'
Joseph John Campbell

LOVE

Humans cannot directly experience the purest form of love, which is GOD's love or GOD-love. GOD-love is not a feeling. It's pure mind. It's a potential existing within nothingness and we cannot feel nothingness. To GOD it's everything. To us it's nothing. That's a bit of a problem when the illusion was co-created by GOD so She could observe all love from all beings from all time.

But humans *can* experience the vibrational version of GOD-love, which is joy. GOD-love is pure potential. That same potential exists within us, just at a human-level, set by our memory-of-potential. We can't feel pure potential since it's nothing. But we can feel potential in motion. The potential behind our thoughts creates waves of vibration (ripples in the pond) and becomes potential in motion or e-motion. We humans can *feel* e-motion. E-motion is potential in motion. Potential is GOD-love. Therefore joy is the feeling created in the body from GOD-love-in-motion, flooding the body in waves of electromagnetic vibration. Ponder that!

Strangely enough, GOD doesn't observe the joy *or* the love we *feel*. GOD doesn't feel vibration because vibration is the opposite of nothingness and GOD is nothingness. GOD observes the powerful and wise *potential behind* our thoughts (stored in our memory-of-potential) that created the love or the joy and then processes *that* potential energy as pure-mind (pure consciousness). This is GOD's language of love - potential.

More precisely, GOD observes all love from all beings from all time by collectively observing all memory-of-potentials from all beings from all time (the Divine). GOD becomes the Grand Observer of the Divine, or G.O.D.

HAPPINESS

The process of joy always starts with happiness. The EGO deranges the potential balance of the mindset. You partially embrace an unbalanced mix of potential-aspects to feel part-joy which is happiness. You may also feel the occasional moment of joy. To stay happy you have to keep making the conscious choice to respond to life with the sort of positive thoughts on page 241, hence the practice of OPTA.

The practice of OPTA slowly equalises the potential-aspects within each mind to slowly balance the mindset. You'll feel a shifting equilibrium of *both* desire and inspiration and your happiness will slowly start to shift into joy. If you keep practising OPTA, your mindset will continue to balance with even higher levels of potential and you will feel more and more moments of joy. Sure, you will have good days and bad days, no one's perfect, but the more you practise OPTA, the more you will overpower your old EGOic habits of thought, the more joy you will feel and the more you will en-joy yourself.

You have to fill yourself up with joy *first*, like a water tank, by re-membering as much powerful and wise potential as you can. That powerful and wise potential will urge powerful and wise thought-action, which will fill you with powerful-desire and inspirational-wisdom and create the feeling of joy within you. That's what is meant by the term 'happy *within*', where 'within' refers to the powerful and wise potential stored in your mind, which creates the feeling of joy *within* you.

Once you are happy *within* then it doesn't matter what you do, you'll always be happy. No matter what kind of job, relationship, place, person or thing, you have the tools within your mind to respond to life with powerful and wise thought and feel happy (joyful).

For the same reason, your purpose is NEVER to make other people happy … ever. Only *they* can do that themselves by embracing their own power, wisdom and joy, *within* their own mind. You can NEVER do that for them because you have no more access to their mind and what they think about, than they have access to your mind and what you think about. You can help them in a state of joy but you cannot make them happy

SERVICE

Joy is all about filling yourself up with joy, *first*. Some people consider this to be selfish, but it's being true-to-self and that's your birthright - to en-joy yourself. If that makes you a 'selfish person' then I say, let the entire world be selfish!

Only after you have filled yourself with joy can you share that joy with others because you cannot give away what you don't have within. Only when you share that joy with others does that joy turn to love. Love is the highest form of human experience. It's a human *being* GODlike.

That's the true meaning of life. But you cannot get there until you en-joy yourself, *first*.

You fill yourself up with joy as you evolve. It usually takes decades to do and lifetimes to master. You can fill yourself up slowly via the snail-like en-Light-enment of necessary-challenges or you can fill yourself up much faster with purposeful-challenges - doing what you love because you know what you love doing.

Once you have filled yourself up with joy you then have something to give to others - potential - and the sharing happens quite naturally. As you fill yourself up with joy, you fill yourself up with potential. As you fill yourself up with potential your increased positivity slowly erodes the EGO. The EGO represents the false-self or the self (in a bad way). It's more concerned with its own subconscious agenda than the needs of others. It's quite self-focused, self-orientated and self-ish.

The more you resolve the EGO, the more you remove the *self* from the picture and the more your life becomes about other people. This is where you naturally begin to share your joy and hence your potential with other people. The more joyful you are the more security, capability, expression and knowing you have within you, which means the more you can *help people* in a secure, capable, expressive and knowing way. This is the principle of service - sharing your power and wisdom in state of joy, helping others.

You can serve others in a myriad of different ways such as your job, daily life and social circle or as a parent. The table below describes some of the more common powerful and wise thought-action-feeling that you could share with others as joyful-service:

Good habits of **SECURITY**	
You	Courage, excitement, strength, life-force
People	Trust, support, willingness
World	Adventure, travel, safety, security
Things	Wealth, abundance, generosity

Good habits of **CAPABILITY**	
You	Energy, willpower, strength
People	Reliability, trustworthiness, motivation
World	Achievement, success, recognition
Things	Skill, talent, appreciation, versatility

Good habits of **EXPRESSION**	
You	Fulfilment, purpose, importance, emotion
People	Communication, understanding, fairness, personality
World	Change, respect, legacy, expansion
Things	Art, creativity, individuality, contrast

Good habits of **KNOWING**	
You	Decisiveness, intuition, peace, creativity, imagination
People	Fairness, wisdom, equality, respect, compassion
World	Service to the individual, community, humanity
Things	Gratitude, care, respect, environ-mentality

Remember the story, 'I Washed My Hands and Brought Down the Stock Market!'? Remember how the most mundane action can have such a massive effect? When you serve people, without EGO and in a state of joy, it can *change the world,* especially if the service is done willingly and without conditions. Joyful-service, as it's passed from person to person, as it *moves* from person to person, really is potential *in motion.*

LOVE

Service is using your potential strengths to help others but you don't have to be a slave. Joyful service is a choice *you* make. You serve people all day every day, anyway. For example, no matter what type of job you have, you are serving others in some way and there is no better feeling than helping others. It's what keeps most people *in* their job, even if it's boring, unhealthy or soulless. Every human relationship involves some form of service. Whether you choose to serve others in a state of joy by sharing your potential is up to you. Only the EGO will stop you from serving others in a state of joy.

But it's a fairly easy choice to make. The joy in service comes from *you* expressing your powerful and wise potential, which makes you feel good anyway, but when you do so to help others; it actually makes you feel even better. Think about the time you helped someone, without EGO (approval, guilt, obligation, false-humility) and just because you could and it felt good. You always feel great afterwards. It's that warm fuzzy feeling you get when you help an elderly person cross the road. It's the warm fuzzy feeling you get when you help someone because you have the skills and know-how to help someone. It makes you a supportive, kind and generous person.

It's a warm fuzzy feeling because it shifts your feeling of joy into a higher, more blissful state of support, help and care for others. This feeling also inspires other higher feelings such as gratitude and compassion. The combined effect of these feelings is otherwise known as ***love***.

Love is probably the most misunderstood human experience but its primary source comes from helping others in a state of joy *because you can* (have the potential to do so). That doesn't make a very good Hallmark™ greeting card but it's the Truth. Love is what you get when you share your potential and so you can 'love' someone in a myriad of different ways.

The most obvious is a loving relationship but what's every good relationship based on? The answer is trust (security), intimacy (desire), commitment (capability), communication (expression), romance (inspiration), respect (knowing) and having fun together (joy). It adds an interesting and quite a naughty twist to the phrase 'doing what you love because you know what you love doing'. It could well be doing *who* you love because you know *who* you love doing.

Then there's doing what you love because you know what you love doing. If you do that *to help others* then it's the true meaning of the statement, otherwise it would be doing what you *en-joy* because you know what you *en-joy* doing. Doing what you love because you know what you love doing *to help others* includes turning a hobby into a fulltime paid career, parenting (the ultimate form of joyful service), developing a great idea (especially if it serves a genuine human need), volunteering your skills, teaching or sharing your potential with the public via writing (Hi!), music or art.

The feeling of love is the reward for joyful-service. It's the highest form of joy and there is no feeling grander than love. There is no human experience greater than love. It's the closest a human can get to the feeling of being GOD - a human *being* GODlike.

You can see love in all human endeavours that involve joyful-service but love starts at parenting. The constant, self-less, dedicated joyful-service required to raise a child is what creates most of the love on the planet. The love this creates is what initially bonds the parent to the child, even before it's born. The love this creates is *why* parents love their children. Not because they 'love them' but because they used every ounce of their potential to raise that child as best they could (serve them) and that service made them feel love (or at least as happy or as joyful as their EGO allowed them to be). That love was shared with the child to bond parent and child together in a framework of love.

The love that parental service creates is also what shifts the mindsets of young parents from self-orientated individuals into caring, loving people. You can see this effect in any young adult who has recently had a child. The years between 22-28 years are all about the *being* the EGO. There is where you live and breathe your subconscious negativity and that subconscious negativity often makes a person self-orientated, egotistical, nonsensical and spiritually unaware. En-Light-enment shifts people out of this stage between 25-28 years of age and continues from 29-35 years of age (en-Light-enment is a lifetime journey but it *starts* here). It's no coincidence that between 25-35 years is usually when couples have children because the love (and purposeful-challenge) of parenting is often required to kick-start that en-Light-enment.

THE JOURNEY OF LOVE

The journey from happiness, to joy, to love takes decades of en-Light-enment. Most people don't hit the love stage until the mid to late thirties and it doesn't become part of you fully until mature love of the forties and early fifties. The true cycle of self-less loving-service extends from 63-84 years. You master love some time in that cycle, usually in your seventies as a wise elder long after you've left the work force. Your 'job' then, is to lovingly serve your adult children and grandchildren by sharing your lifetime of potential and joy with them.

What *level* of love you master in your seventies depends on what sort of joy and service you practice now. En-joy yourself and love others *now* and you will retire in a state of bliss *then*. Life naturally evolves you to that point anyway, slow like a snail, but if you practise OPTA you can avoid a lot of unnecessary pain and suffering and get there much faster.

People often ask me why I don't write more about love. Love is an outcome. It cannot be taught. Only potential can be taught and that eventually leads to love. Remember at the start of the book I said, 'I am clearly and bluntly going to point the finger straight at 'Thyself', which is you … because it draws *you* into the book and makes it personal'? That's because spirituality is all about you, en-joying yourself, *first*. Otherwise you have nothing to give to others and if you have nothing to give to others then you cannot experience love. To write a book about love is to write a book about power, wisdom and joy. That's how I became a healer and that's my form of healing.

WHAT 'FALLING IN LOVE' REALLY IS

Then there is the issue of falling in love, romance and dating and how that all gets screwed up with joy, true-love, Valentine's Day and Internet dating. Now that's a whole other book! But here goes!

You were once whole (ONE) before you split into the male conscious mind of power and the female subconscious mind of wisdom (TWO). Both men and women have this same split mindset. The male conscious mind is the mind you're more familiar with. It's the more obvious, 'out there', exposed part of the mind that thinks. The female subconscious mind is the more hidden, inner, mysterious mind. There's a legitimate sexual-genital analogy going begging here, but I'll let you work it out for yourself.

Because you have identified with the male-conscious mind, you think it's *you.* It's only half you, being that you are a mindset, and so *you* energetically feel as if there is another part of you (the subconscious mind), *somewhere,* that's hidden or 'missing'. This generates a subtle feeling, deep within you, that you are *incomplete*.

Whether you are male or female, you yearn for your 'missing' female-subconscious mind, or 'missing-mental-partner'. You long for it. You search for it. You crave it. It's this yearning for your missing-mental-partner that forms the basis of sexual attraction, falling in 'love' and what we think is 'love'.

We can't find our own missing-mental-partner so we fall in love with someone else's mental-partner - their subconscious mind. I call this sex-mental-attraction. The same principle applies to homosexual relationships because the sex-mental-attraction is between conscious mind and subconscious mind; not between physical-male and physical-female. Sex-mental-attraction bypasses gender.

You are the sum total of what's stored in your subconscious mind - thought patterns, preferences, belief systems and memories. In short, you *are* your subconscious mind. The conscious mind simply processes those memories, as thought-action-feeling, *now.* So when you meet another person you actually meet their subconscious mind because it 'speaks' to you *now,* through their conscious mind.

When you spend more and more time with the person, so spend more and more time with their subconscious mind, you get to know their subconscious mind. You gradually become influenced by it, familiar with it and tantalised by it. If they consciously exhibit thoughts-actions you like, and the thoughts-actions you lack and so crave, then it's as if you have found your missing-mental-partner.

You then mentally and potentially 'reunite' with your missing mental-partner. Being that the subconscious mind is feminine (the subconscious mind is a female energy in both men and women) you fall in love with 'her'. When they go away, you miss 'her', you long for 'her', you want to reunite with 'her'.

This is all the more complicated with the addition of dinner, wine and sexual urges. Be wary of wine-infused, chocolate underwear! This is all the more karmic with the necessary and purposeful challenges of pre-sent relationships. It's complicated!

Either way, when you fall in love with their subconscious mind it becomes your external, surrogate, mental-partner. You create the illusion of being 'whole' again or wholly (Holy) again. That Holy feeling combined with sexual attraction and passion is what humans call 'love'.

Considering that love is fuelled by a male-conscious mind/female-subconscious mind attraction and that mindset is an electromagnetic system, the poles of which are potentially reversed between male and female, then attraction between male and female *really is* magnetic (wisdom) and we *really do* turn each other 'on' (power).

SEXUAL URGES

The primal sexual urge within us that's built into the power of desire is actually a physical expression of the conscious mind yearning to reunite with the missing-subconscious mind and be whole again. This yearning continues long after you enter into a relationship with another person, because your mental-partner is still 'missing' and so the yearning is always pre-sent. The same desire underpins the need for affection, companionship and love, which flows onto the need for friendship and approval.

For men, this feeling is far more intense because men are more aligned with the conscious mind of power. The missing-mental-partner is the subconscious mind, which is a feminine energy and so a man's urge to find and reunite with his missing-*female*-mental-partner is far more intense. Consequently, men's sexual urges are far more intense.

Women are more aligned with the subconscious mind of wisdom. Their missing-mental-partner is already part of their alignment. It's still 'missing' but women feel the separation less than men. Consequently, women's sexual urges are less intense.

By comparison, when it comes to *primal* sexual urges, women can take it or leave it more than men, whereas men have two heads; the smaller one drains the blood out of the bigger one so the bigger one can't think of a reason why it shouldn't ask for sex.

TRUE LOVE

What does this mean? Did we just marry and have babies with an external subconscious surrogate? Is love wrong? Have we all totally missed the idea of true love? The answer is of course, 'No'. Love is grand!

The ever pre-sent urge of sex-mental attraction is what brings two people together and helps *keep* them together, in what *eventually* ends up being the greatest of all human challenges - intimate relationships. Without it, there would be no relationships.

After the initial buzz of a new relationship subsides (usually signalled by the first person willing to fart on the other person's leg in bed), the relationship becomes a challenge. Some say 'hard work', but I prefer the word 'challenge'. Intimate relationships are always challenging - even the good ones. They're often a very purposeful-challenge and a whole bunch of necessary-challenges all rolled into one. They can be joyful and hurtful, loving and sad, rewarding and disappointing ... but always challenging.

Intimate relationships challenge you on a deep, personal, intimate level and uncover a *huge amount* of potential. That constant en-Light-enment very quickly shifts that relationship from a happy-exciting relationship into a deep-joyful relationship, then into joyful service, then love and finally into true-love.

Relationships are always potentially reciprocal. You potentially grow together and so you always get back what you give. The more you en-joy yourself the more you will be en-joyed. The more you serve, the more you will be served. The more you love, the more you will be loved.

The added bonus is that all the potential from all the challenging experiences that led to the true-love, get stored in the subconscious mind. This is the missing-mental-partner that you initially fell in 'love' with, yet every day that missing-mental-partner becomes more evolved - more powerful, desirable, wise, inspirational, joyful and loving. You both get to fall in love with your missing-mental-partner, all over again, every day, 'till death do us part' - except this time the love is real because the joy is real, the service is genuine and so the love is real. This is why true-love can only come from a challenging long-term relationship.

THE TRAP

The trap is that you have to en-joy yourself *first,* separately as a single person, before you can find true-love. Without your own potential water-tank, you not only have nothing to give in the initial stages of the relationship, you won't be able to experience service or love either.

More importantly, you won't be happy *within* yourself. If you're not happy, then you'll rely on your partner to *make you* happy. This NEVER works in a relationship because your partner CANNOT make you happy. Happiness can only come from within *your own mind,* via your own powerful and wise thought. If you don't go within, you go without.

After the initial Holy excitement of sex-mental-attraction fades (and it will because it's a façade), your external source of happiness will also fade. It cannot last because it's not true happiness. You become unhappy. If you're no longer happy in the relationship because your partner no longer 'makes you happy' then undoubtedly your partner's no longer happy either. The relationship breaks down.

Unhappiness always activates the EGO. You then react badly to each other. The negativity of the EGO makes a bad relationship worse - it feeds on itself. Thoughts of insecurity create doubt and mistrust. Thoughts of failure create a lack of will and determination to fix the relationship. Thoughts of depression create a lack of communication and the problems fester. Thoughts of ignorance create a lack of understanding and intimacy. Combine all this with apathy, boredom and all the associated feelings of seriousness and you end up in a serious relationship.

Serious relationships tend to sweep a whole lot of anger and hurt under the carpet. This creates a lump of negativity that grows and grows. Sooner or later that lump grows to such a height that the relationship cannot move past. There is just too much hurt, too much anger, even too much hate (the first thing the EGO does is blame the other person for the unhappiness. This causes resentment and hate). The relationship becomes permanently damaged and fails.

But it's not all 'bad' (even though there is no bad). It may not feel like it at the time, especially if your divorce lawyer is now having guilt free, rich-person-sex with your wife/husband. A broken marriage or a bad relationship will uncover *huge* amounts of potential in both of you and your children. The purposeful-challenge organised by you then becomes a necessary-challenge organised by YOU.

You could even turn your bad relationship into a purposeful-challenge by either trying to fix the relationship or by leaving the relationship with power, wisdom and joy.

THE SEVEN YEAR ITCH

Like all other major challenges in life, falling in love happens in cycles of seven. The purpose of each seven cycle is to challenge both of you, purposefully or necessarily, on a higher level, to make sure you evolve as individuals and as a couple. This is a big step. Your Soul knows this and so YOU, 'asks you' (subconsciously), whether you wish to proceed to the next level. This subconscious, 'are you sure?' re-evaluation is often called 'the seven-year itch'.

It's the challenge of the relationship that's most important, yet it's the challenge of the relationship that most people complain about as being 'hard work'. Unfortunately, with many people living in a state of EGOic life-laziness, many couples aren't willing to face the challenge (good or bad) and so the subconscious re-evaluation often ends a relationship. This is why one in every three marriages ends in divorce, usually after (or close to) a seven-year 'itch', a fourteen-year 'itch' or a twenty-one-year 'itch'.

The answer: face the challenge. Try and better your relationship by using that relationship to practise OPTA. Respond to everything with OPTA. That's the challenge.

Lead by example. Be the positive person. Your partner will either resonate with and join you (eventually), or it will really piss him/her off and he/she will leave.

You are NOT responsible for your partner's happiness ... ever. Your only responsibility is your own joy. That's your duty therefore OPTA is your obligation. It's all you can do.

SUMMARY

- Humans can't experience pure love or GOD-love.
- Humans can experience a vibrational version of that God-love called joy.
- We can't feel potential because it's nothing but we can feel potential-in-motion which is e-motion.
- GOD observes the powerful and wise *potential behind* our thoughts (stored in our memory-of-potential) and then processes that potential energy as GOD's language of love - potential.
- Joy starts with happiness.
- Once you are happy within, then it doesn't matter what you do, you'll always be happy.
- You CANNOT make another person happy.
- Happiness will balance the mindset so you can feel joy.
- You have to fill yourself up with joy first otherwise you have nothing to give.
- As you fill up with joy, your EGO dissolves and life becomes less about the self and more about other people.
- This shifts your joy from being all about you, to being more about other people.

- You naturally share you joy and hence your potential with others to help people in a state of joy.
- This shifts your joy into service.
- Serving others in a state of joy is the basis of love.
- Love is the grandest of all human experiences.
- You can love people in a myriad of ways such as being a parent, partner, friend, etc.
- Doing what you love because you know what you love doing, *to help people*, is the most en-joyable way to serve others and includes your career, hobby, teaching, art, music, etc.
- The feeling of love is the reward for joyful service.
- It takes decades to master love.
- Falling in 'love' happens when your conscious mind unites with the subconscious mind of another person, reuniting your conscious mind with its 'missing' mental-partner.
- This sex-mental attraction continues long after you have entered a relationship.
- Women should give men more sex (OK, just kidding! Just seeing if you're paying attention. Although it is a nice idea!).
- Intimate relationships are very challenging and uncover a *huge amount* of potential.
- True-love happens when the potential of that challenge evolves the relationship from happiness, into joy, into service, into love and finally into true-love.
- You both get to fall in love with your missing-mental-partner, all over again, every day, 'till death do us part'.
- You have to be happy, first, as an individual in a relationship otherwise you have nothing to give.

- Unhappiness activates the EGO.
- The EGO makes you both react badly and the relationship can quickly become a seriously bad relationship - challenging but unlikely to last.
- The seven-year itch corresponds with the seven-year cycle of challenge.
- Unfortunately, many couples aren't willing to face the challenge (good or bad) and so the relationship ends, usually after (or close to) a seven-year 'itch', a fourteen-year 'itch' or a twenty-one-year 'itch'.
- You are NOT responsible for your partner's happiness … ever.
- Your only responsibility is your own happiness; it is your duty.
- Live your dreams.

WHAT'S GOING TO HAPPEN IN 2012?

Have you noticed the world is changing at a rapid rate? Have you noticed the world is suddenly embracing potential? Just one example of this is reflected in the recent increase of prime-time television shows such as celebrity chefs, talent shows, dance shows, and 'How To' programs. Not only are these TV shows all about challenging and uncovering potential but they also reflect an ever increasing awareness of self development and positivity. This is the return of the Light.

Without getting into a religious debate, the 'return of Jesus' that Christians all get very excited about is not the return of a man, it's the *return of the Light* and Light comes from power and wisdom. So the 'second coming' is the return of power, wisdom, joy and love to this planet, *en masse*. It's not the return of Christ. It's the return of Christ *consciousness*.

We're heading into an exciting new era of human evolution where the entire human race will collectively embrace powerful and wise potential, producing a parabolic increase in joy and love across the planet. This is what I call The Potential Age.

When will this happen? According to the United States Census Bureau, annual births worldwide are expected to remain constant at their 2011 level of 134 million people per year. That's approximately 15,000 babies being born every hour, 254 babies a minute or just over four babies a second! That's a lot of mind being added to the planet. The common metaphysical theory about The Potential Age is that when the amount of potential stored within the minds *of all people* reaches a certain *collective* level that collective potential will influence the brains of *all people* to think with power and wisdom. This is based on the 100th Monkey Syndrome.

THE 100TH MONKEY SYNDROME

In 1952, on the Japanese island of Koshima, scientists devised an experiment to study the behaviour of monkeys by dropping sweet potatoes in the sand. The monkeys liked the taste of the potatoes but not the sand. One day, a monkey named 'Imo' washed the potatoes in a nearby stream. She taught the trick to her mother and her playmates, who taught it to various other monkeys.

Apparently, ninety-nine monkeys learned to wash their sweet potatoes between 1952 and 1958. One day, the one hundredth monkey learned this skill. Suddenly and miraculously, all the other monkeys on the island began to wash their potatoes in the stream before eating them. Even more amazing, scientists observed that monkeys on the other islands, as far as five hundred miles away, also suddenly and miraculously started washing their sweet potatoes before eating them!

This prompted the theory that our minds must somehow be connected, as *one,* creating one gigantic, collective, connected subconscious mind (the collective consciousness), or collective-mindset. The potential energy within the minds of all people accumulates in the collective-mindset, to a level that eventually and collectively influences the brains of *all people,* to think in a similar way. That new way of thinking then gets stored in the subconscious mind of each individual, adding to the collective-mindset, strengthening and reinforcing that new way of thinking. The net effect is a *parabolic increase* in a certain type of thinking.

The same thing is happening to the human race *now,* with the potentials of wisdom and power. If you consider the 100th Monkey Syndrome in conjunction with how fast the Internet and modern media can spread new ideas and ways of thinking, then we're in for a sudden change, the likes of which the human race has never before seen. It will bring tears of joy to your eyes, literally, and it's already started. It's a great time to be alive!

2012

Within the Mayan culture that lived hundreds of years ago, was a group of Galactic time-keepers with an intimate knowledge of the cycles that occur within the illusion of space-time. These Mayans were Ascended Masters and according to their Galactic calendar, the day the 100th Monkey Syndrome reaches its tipping point is December 21, 2012. From that day forth, the potential within the collective-mindset will reach a level that will influence the brains of all humans to think with more security, capability, expression and knowing.

So what's actually going to happen on December 21, 2012? Nothing! The year will continue as normal and progress into New Year's Eve just like every other year. People will party the night away, eat too much, drink too much, kiss a whole bunch of people they don't know and then go to bed reminiscing about the year of 2012, just like they do every other New Year's Eve.

On January 1, 2013, every human-being on the planet will wake up with a new level of potential, gained from the shift on December 21, 2012. But that new level will be small. It's a parabolic shift and New Year's Day represents the bottom of the curve. In addition, the shift itself is subconscious, so most people won't even notice the small increase in potential. They will celebrate New Year's Day, sleep in, go to work or whatever they do ... but they *will* be different.

INDIVIDUAL CHANGE

From December 21 onwards, the powerful and wise potential within the collective-mindset will increase. As every human slowly embraces this new level of potential, their thoughts will slowly change. Old patterns of insecurity, failure, depression and ignorance will slowly be replaced with new thoughts of security, capability, expression and knowing. Seriousness will slowly be replaced with joy. Hate will slowly be replaced with love.

The powerful and wise potential behind these thoughts will be added to each person's subconscious mind, adding to and strengthening the effect of the collective-mindset, which then urges *more* positive thoughts, which then adds even more potential to the collective-mindset and so on. It's truly parabolic and the human race will change.

For example, people will find the power to conquer their fears and take a chance on life. They will release the material bonds of money and possessions and follow their desires. They will create an abundance of everything. They will suddenly feel the urge to get healthy and take better care of themselves. They will feel better about their body image and discover a new level of sexual intimacy. They will have new energy levels and willpower. They will learn new skills, finish tasks and achieve goals. They will be more successful.

They will express their unique personalities to the world. They will stand-up for themselves and their point of view. They will feel a new sense of purpose and direction. They will make great decisions based on great ideas and vision. They will experience a new clarity about their life's purpose. They will embrace change. They will forgive. They will let go of their guilty past and stop worrying about the future.

They will focus on what they like and learn from what they don't like. They will embrace *now,* and be more connected to the wonder and beauty of life. They will seek spiritual knowledge as they re-member the Truth. They will learn things in months that previously took years. They will become self-aware in a fraction of the time and with very little effort. They will reconnect with Mother Earth. They will reconnect with their Soul.

Their lives will self-focus around joy. They will suddenly realise that life is not about money, possessions or work and that true success can ONLY be measured by how much joy they feel. They will en-joy themselves more. They will en-joy others more. They will en-joy life more. They will feel a powerful-desire and inspirational-wisdom to do what they love because they know what they love doing. They will feel a powerful and wise urge to serve others, in a way that best expresses their own potential. They will feel love on a level previously unimaginable.

COLLECTIVE CHANGE

But that's only on an individual level. The added bonus of The Potential Age is that a collective increase in potential means a collective increase in awareness. The human race will suddenly wake up and then take massive action!

Take the environment for example. Humans will soon collectively realise that Mother Earth is a living being. She is alive and She thinks. She is beautiful, peaceful and GODlike. In comparison, the human race is ugly, dirty and arrogant and treats Mother Earth with such disrespect. Humans grab, hoard and plunder Her resources and then proudly beat their chests about modern life and what they have achieved.

Humans will soon realise that their fragile, fleshy-bag-of-water-bodies run on fresh air, water and living-energy-food (fresh and raw, fruit, vegetables, nuts and seeds). Yet, they have successfully ruined the atmosphere, polluted the water systems and processed the life-force out of most food. In addition, 'the powers that be' are *still* fiddling with genetics, bio-chemical weaponry, nuclear nastiness and GOD knows what other kinds of environmental suicide.

Humans will soon realise that the last 300 years of factory automation and mass production since the Industrial Revolution, has simply been a massive, voluntary journey into the impurity of pollution, extinction and depletion. As a race, humans have 'reacted badly' and gone downhill and it's now time to 'cheer up' by voluntarily choosing to take a massive voluntary journey back to the purity of nature, life and sensible-abundance.

And to do so, the human race will have to soon embrace their marvellous new levels of potential and take *massive action* to clean up the dreadful mess they've made of the planet. That's our new purposeful-challenge as a race of beings and it will uncover massive amounts of new potential.

This new purposeful-challenge will also produce an unbelievable change in lifestyle. Humans will soon discover unlimited one hundred per cent clean energy. The pollution created by coal, petrol and diesel will cease. There will be more raw food production and better health. Energetic balancing will replace modern medicine. There will be fewer chemicals in products and more natural organic alternatives. There will be new industries and new opportunities for all, including a more compassionate distribution of wealth.

That's the future of this planet. That's what's going to happen in 2012 and beyond. Not death, war, carnage and disease but power, wisdom, joy and love. Search deep inside your heart and you will know this to be true. We cannot avoid it. We cannot get it wrong. We cannot fail. Our journey back to purity is *guaranteed* and everything is going as planned - since the illusion is engineered by GOD and facilitated by Gods.

There will still be extreme-duality - that never changes because humans need the contrast, otherwise we wouldn't know, nor would we seek. But the entire system of extreme-duality will shift up the scale, so that our worst will be replaced by our previous best and our new best will be at a level never before experienced. The distance between the poles of extreme-duality will stay the same (or at least similar), only the poles will have shifted upward, *as one.* You get the same effect if you shift a set of soccer goals from a B-Grade game over to an A-Grade game. The gap between the goals is the same, but the potential is higher. It's a great time to be alive.

THE CHANGE WILL BE SLOW

The change of The Potential Age will be parabolic but the initial stages of change won't be sudden. A parabolic increase is gentle, curvaceous and effortless and in the beginning, slow and gradual. This will allow people to initially change at their own pace, in their own time and in their own way. Then as the change shifts parabolically into a faster curve, the change will still be graceful, effortless and amazing for *everybody*, because *everybody* will be doing it (mostly).

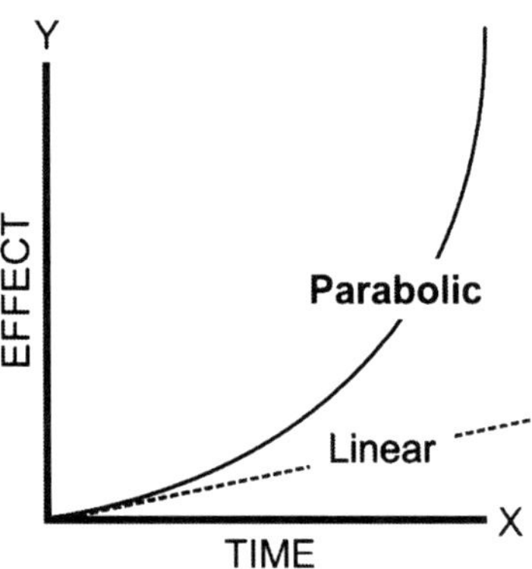

(Figure). A parabolic change (effect over time) is curvaceous - gentle at the start with a sudden surge towards the end.

My guess is that the integration of this potential will follow the 3 x 7 = 21 year cycle of sevens. Integration of this new collective potential will follow the following table:

1	2012-2019	1st major cycle	1-7 years	Develop
2	2020-2026	1st major cycle	8-14 years	Strengthen
3	2027-2033	1st major cycle	15-21 years	Reinforce
4	2034-2040	2nd major cycle	22-28 years	Develop
5	**2041-2047**	**2nd major cycle**	**29-35 years**	**Strengthen**
6	2048-2054	2nd major cycle	36-42 years	Reinforce
7	2055-2061	3rd major cycle	43-49 years	Develop
8	2062-2068	3rd major cycle	50-56 years	Strengthen
9	2069-2075	3rd major cycle	57-63 years	Reinforce

(Table) The integration of potential, following a (3 x 7 = 21) year cycle of sevens

It will probably take the first major cycle (3 x 7 = 21 years) to accomplish a *noticeable* integration of this new potential, mainly because it takes three cycles of seven years to accomplish a noticeable integration of everything else. There will be a further twenty-one-year integration of that potential before it becomes a *very noticeable* integration of potential, *collectively*. In short, 2054. After that, I have no idea what's going to happen. Suffice to say it will be fun!

The same thing happened with the environment. It took about twenty-one years and a lot of hard work by thousands of people to make the environment an 'issue'. This got millions of people thinking. It then took a further twenty-one years for the potential behind those environmental thoughts to accumulate in the collective-mindset at a level powerful enough to urge all minds of all people to think environmentally. Suddenly, almost overnight, an entire race of westerner woke up with a strong environmental *urge*.

Some embraced it, others ignored it but the urge was always there for all. It didn't take long before a parabolic increase in an environmental type of thinking became mainstream in western thinking. Now, topics such as global warming, recycling, de-forestation and pollution are an integral part of politics, world news and daily conversation.

I can actually remember the year that environmental issues went collective (for me anyway). It was 1995 and it seemed like on Monday, I didn't really care that much about the environment, whereas Tuesday, I woke up as a 'greenie'. I got out of bed, angry with the way the Earth has been treated. I was suddenly worried about the ozone layer. I wanted to swim with a dolphin. I wanted to plant a tree. I recycled my first cardboard container. I can even remember thinking to myself, 'Wow, that's weird. When did I become a greenie?'

The same thing is happening right now with spiritualty. Spirituality (personal growth, positivity, self-help and new age thinking) first became popular in western society in the early 1970s and was slowly integrated from 1970-1991. It then became an issue from 1991-2012, following a lot a very hard work from people like Louise Hay, Tony Robbins, Deepak Chopra and Oprah Winfrey. On December 21, 2012, spirituality will become mainstream (potentially collective). From that day forwards, the entire human race will feel an *urge* to be more spiritual, which can only come from being more secure, capable, expressive, knowing, joyful and loving - since that's what spirituality *is*.

But don't be disheartened. What I am referring to is a an *entire race* of humans, collectively embracing a new level of powerful and wise potential so that the *entire world* thinks and acts with security, capability, expression and knowing to feel constant joy and love. That's going to take a few decades, if not 63 years, or maybe 84 or 105 years. Who knows?

Imagine what that's going to be like. Think about entire communities, cultures and countries embracing power, wisdom, joy and love. Image the world our kids will inherit when that type of thought-action-feeling becomes part of everyday language, filtering into government policy, education and the workplace. Imagine a massive and collective resolution of the human EGO ending all wars, poverty, crime and hate.

But that's the general population. You're already on a spiritual path. If you're willing to fully embrace the powerful and wise potential of The Potential Age, to do what you love because you know what you love doing, in order to serve others in a state of joy and love, then you will experience joy, love and en-Light-enment that's *way beyond* anything experienced by the general population. It's already happening to you, now.

THE JESUS-LIKE STATE

Within the first 21-42 years, I fully expect a few people on the spiritual path to reach a Jesus-like state. This is a state where you embrace *so much* powerful and wise potential that no matter what happens (around you, or to you), you *always* respond and react to life by consciously choosing powerful and wise thought-action, to maintain a *constant* state of joy, service and love. That's a human *being* Jesus-like. This is also the premise behind the Christian doctrine 'turn the other cheek':

> *'If someone strikes you on the right cheek, turn to him the other also. And if someone wants to sue you and take your tunic, let him have your cloak as well. If someone forces you to go one mile, go with him two miles. Give to the one who asks you, and do not turn away from the one who wants to borrow from you.'*
>
> Matthew 5:38-42

In more modern kiddie lingo, if someone's slapping your face, suing you for everything you've got, dragging you all over the place against your will, bleeding you dry for cash, borrowing everything in the world you value as sacred yet never returning it or returning it broken … and *yet* you're still grinning with joy and still willing to serve them with love … then you're Jesus-like. You have so much personal power (self-control, courage, will-power, etc.) and so much personal wisdom (compassion, patience, understanding, etc.), that you're literally being a human *being* Jesus-like. You have become OPTAmised.

If you're like the rest of us and you've slapped them back, counter-sued them, told them to %$#@ off and then raided their house in the middle of the night to get back all your stuff, then believe it or not you are being a human *being* impure. You will reincarnate, lifetime after lifetime, here on Earth, as a human, until you re-member enough potential so that you can react to such a situation with pure power, wisdom joy and love. That's the challenge!

Someone will achieve that state for sure - possibly many - in the first few decades of The Potential Age. That person will not only be a 'swell guy/girl' but their OPTAmised level of potential will co-create events into their *you*niverse that will seem nothing short of *miraculous* - a human *being* Jesus-like.

You may not see it on YouTube™ because the Jesus-like people will most probably be leading by example, humbly in the background, but those types of people will think and do Jesus-like things and it will mark 'The Second Coming'.

I'm quietly confident the Jesus-like individuals will come from the new generation being born now, rather than people who are already here. The people who are here now are already potentially challenged. The new generation of babies being born now are being born with staggering levels of potential. It's like they've all been dipped in 'Dalai Lama juice' and then rolled in 'Richard Branson crumbs'. Not only that, they're being born into a parabolic shift in collective-potential. They will start life, potentially, where most adults finish.

These children will evolve at a level previously unattainable (by most). They will grow up to be our greatest teachers of joy and love. Children will *teach us* - not the other way around. Children are already displaying a level of joy beyond most EGOicly limited adults. They are the masters of joy. Joy *is* child's play, not adult seriousness. Heck! Most people's pet dogs display more joy (and love) than most adults do - sad but true - whilst doctors marvel at the 'mysterious' healing properties of Pet Therapy - derr! But place these joyful children into The Potential-Age and they will soar, beyond most adults.

As children they will embrace the new collective-potential and lead by example by *being* childishly joyful. They will also embrace the new technologies as each younger generation always does and they will spread that joy across the planet 'Facebook™' style. As they grow into young adults, and learn to serve others with love, they will express joy and love with a powerful-grace and a humble-wisdom, not *ever* collectively seen before.

They have all the ingredients required to achieve Jesus-like en-Light-enment. That's why they have incarnated here, now. As Souls, they decided to take on the purposeful-challenge of trying to *be* Jesus-like, in their lifetime.

THE SECOND COMING

'Hey! Jesus has been mentioned more than just a few times. What's going on? It's starting to sound like a religious thang!' OK, calm down. Remember, I am not a Christian. I am not affiliated with any religion or spiritual group and I still haven't found a gold-cross-white-tunic combination that matches my rose petals.

In saying *all of that,* I would like to talk about Jesus … a bit. Jesus was a man who incarnated with so much potential that he changed human history. His role was to be a living breathing example of the powerful and wise potential. He was 'the Light' and 'The Way' (the way of thinking).

Apparently, Jesus started his ministry at age thirty, which lasted for three and a half years. That means he died when he was thirty-three years of age. That thirty-three age number wasn't recorded in historical documents just so Pope Gregory XIII could work out what year it was on his now famous Gregorian calendar, based on the number of years since the epoch of Christ (Anno Domini, or A.D.). That age was recorded for a reason - a reason engineered by GOD and facilitated by Gods, as part of the illusion, where *nothing* is coincidental.

Jesus' crucifixion-age of thirty-three was recorded in history to help show humans that en-Light-enment begins between 29-35 years of age. This marks the death (your crucifixion) of your EGO and the beginning of your en-Light-enment into OPTA (your ascension into Heaven). You then start a slow journey back to purity as your mindset evolves from 35-84 and beyond.

You eventually die but you're born again as you're reincarnated into a new lifetime (this is your resurrection), being urged by the potential uncovered in your previous lifetime, via your new lifetime's memory-of-potential. You then start the cycle all over again - born pure, learn the EGO, resolve the EGO, attain en-Light-enment (for that lifetime) and then death, reincarnation, born pure ....

You do this lifetime after lifetime until you become Jesus-like (this is 'Heaven on Earth' or 'Thy Kingdom come, Thy will be done, on Earth as it is in Heaven'). Once you become Jesus-like, you no longer need to reincarnate here on Earth, as a human *being* Jesus-like. You will start a new journey in a Light dimension as a Light-Being *being* GODlike.

When the generation of babies being born today reaches 29-35 years of age (2041-2047) they will begin *their* en-Light-enment. Considering that Jesus reached his maximum potential at thirty-three years of age, that Buddha attained Enlightenment at age thirty-five after meditating for forty-nine days under the Bodhi tree, that a major en-Light-enment cycle happens in humans from 29-35 years and there will be a massive parabolic increase in powerful and wise potential; it's quite possible that some (if not many) of the new generation will reach the Jesus-like state between 2041-2047.

If that idea just seems too touchy-feeling-hippy-tree-huggin'-nonsensical-whoopy-foofness then consider this: a lot can happen in 29-35 years. Think back to what life was like 29-35 years ago! If the current year is 2012, that would make it 1977-1984:

- The original 'Star Wars' movie was released (1977).
- Elvis Presley died (1977).
- The album 'Rumours' by Fleetwood Mac won a Grammy Award (1978) - (damn good album!).
- Kate Hudson was born (1979).
- Ronald Reagan was elected President of the USA (1980).
- John Lennon was shot dead in New York City (1980).
- 'Pac-Man fever' swept the world via the arcade game Pac-Man™ (1981).
- The Commodore 64™ home computer was released with 64 KB of RAM (1982).
- The CD was introduced as a replacement for vinyl records (1982).
- The first hand-held mobile phone (Motorola™ DynaTAC 8000X) was released (1983).

- 'Crack' cocaine developed in the Bahamas, first appeared in the United States (1983).
- President Ronald Reagan re-elected! (1984).
- Apple™ introduced its first personal computer (1984).

(Figure) The Motorola DynaTAC 8000X mobile phone, released in 1983. Photo ©Rico Shen. Wikipedia

Just look at that phone! Look at it! Now glance over at your Android™ smart-phone (or if you're more savvy and have more disposable income, your Apple™ iPhone™). Contemplate the massive leap forward of technological potential. Now imagine an even greater leap forward in human potential, during The Potential Age. A Jesus-like state is very possible. If it doesn't happen in between 2041-2047, then it's most certainly going to occur in the third cycle of The Potential Age between 2055-2075. That's when the new generation being born *now* will be at the right age (71-77 years) to be Jesus-like elders, *then*. We'll have to wait and see.

Depending on your age and evolution, you *may* be able to achieve such a Jesus-like state, if you're willing to fully embrace the new collective potentials of The Potential Age and practise OPTA. If you don't, then don't panic. You'll be back, lifetime after lifetime until you do. It's more likely, in this lifetime that you will achieve moments of Jesus-likeness.

Either way, it's onwards and upwards from 2013 and life will get better and better, very quickly. We will NEVER go back to the potential we had before, just as we have not gone back to the potential we had in the Dark Ages of Europe, or the world wars or even the tragic music scene of the 1980's. It's a great time to be alive.

And you will know when The Potential Age starts because you'll hear all about it on Facebook, Twitter and Google+. You'll see the effects on TV. Your friends will text positive affirmations to you on their smart phones. It's actually the 'sneaky Soulful' reason *why* we have satellite communication, mobile phones and the Internet.

Don't you find it peculiar that the human race has had such a massive leap forward in global communication technology? In the last hundred years we have gone from the wireless telegraph to a worldwide network of instant communication. This was Divinely inspired and then

Soulfully facilitated to install a communication network across the planet, to help spread the new ways of thinking of The Potential-Age.

The World Wide Web (www) was introduced in 1991. That's twenty-one years ago and twenty-one years is no coincidence (everything happens in cycles of 3x 7 = 21 years and it takes twenty-one years for something to become an 'issue'). The Internet becomes an issue on December 21, 2012. In the next twenty-one years, it will then become globally collective at the exact same rate the human mind becomes potentially-collective. As our subconscious minds join together in the collective-mindset as a gigantic web of potential, our computers will join together as a gigantic web of communication. The collective-mindset will urge powerful and wise thought-action and the collective-computer-set will spread that thought-action across the globe. We will come together as an online human *collective.* Millions of people will be introduced to personal growth, self-help, positivity, general awareness and spirituality. It's a great time to be alive.

What's really interesting is that the fundamental source of all modern communication technology is the integrated circuit (also referred to as IC, silicon chip, microchip or computer chip). It's what powers all computers, phones, TV and electronics. The integrated circuit is made from silica or sand, which is turned into silicon crystals. This means that all modern communication technology runs on a framework of *crystal power*. It's Atlantis, all over again.

DIS-EASE: THE SIDE EFFECT OF 2012

The world is going to change at a rapid pace and that will create some discord, the major side-effect being self inflicted disease. The entire collective-mindset of the planet is going to awaken simultaneously. That means higher and higher levels of potential *for all humans*. There is no escaping this potential and there is no exception. All humans *must embrace* this new potential or it will make them unwell.

Potential translates to positive thought, which creates high frequency vibration. The EGO translates to negative thought, which creates low frequency vibration. Low frequency vibration not only clashes with the high vibrational frequency of the Cosmos (that's why it feels bad) but smothers your cells in a toxic vibrational environment that creates dis-ease.

The word 'dis-ease' is deliberately written this way to help remind people that negative thinking is 'not being at ease with one's thoughts'. If your EGO tries to *resist* the new levels of Light, and reacts with negative thought, it will create dis-ease (feel bad). If left unchecked that dis-ease will eventually become a physical disease. If you still refuse to change and instead cling onto old habits of the EGO, the disease could escalate and become serious or life-threatening. That dis-ease will then become a necessary-challenge, challenging you at a drastic level that 'forces' you to change.

That necessary-challenge is already happening in the Western world, now. Why do you think there has been an incredible escalation in life-threatening conditions such as diabetes, heart dis-ease, obesity and cancer? Similarly, the precursors to dis-ease have also increased including deep-depression, migraines, neck and back pain and panic attacks?

If you still refuse to embrace the new levels of potential in 2012 and beyond then that dis-ease could escalate even more and ultimately kill you - emotional suicide. This is evolution's way of weeding out the potentially weak and ensuring the potentially strong interbreed to ensure each new generation of The Potential Age evolves. It may seem harsh, but it's simply Nature's way. You live in Nature. If you don't like the ramifications of dis-ease then change the way you think. If you refuse to change then *you* must take full (and final) responsibility for your thoughts and actions!

Consequently, now is not the time to be stubborn or push against the flow of life. Now is *not* the time to be life-lazy, by playing it safe in your old unchallenging habits. Now is not the time to harbour the insecurity, failure, depression or ignorance of the EGO. Now is the time to 'get off your arse' and embrace the powerful-desire and inspirational-wisdom of doing what you love because you know what you love doing. Be the best you can be (or at least try to be better) at whatever challenges you the most, *and* gives you the most joy. Take a chance, be a better parent, get healthy, achieve your goals, try something new, etc. Live your dreams.

Push yourself! If you feel fear or lacking, find the security-potential within you to act with courage and abundance. If you fail or can't be bothered, find the capability-potential within you to act with success and willpower. If your expressions are suppressed or devalued, find the expression-potential within you to stand up for yourself with a new purpose. When you're confused or don't know, find the knowing-potential within you to make a decision and find out.

En-joy yourself then share that joy and the power and wisdom behind that joy, by serving others in a state of joy. You'll be loved for it (literally) and you'll love it (literally). That's a pretty easy choice to make *really*, when you consider the slow, painful, deathly consequences if you don't!

THE END OF THE WORLD?

What about 2012? Isn't the world supposed to end on December 21, 2012? Firstly, if you're reading this book and it's after 2012, please read on because this next section explains what happens *after* 2012, which is probably more important to know. Secondly and most importantly, the Mayan calendar date of December 21, 2012 predicts the end of a cycle and the start of a new cycle. That's it. It *never* had *anything* to do with the end of the world. It's the same cycle the Australian Aboriginals, the Egyptians, the Hopi Indians, the Chinese Sages and the Indian Mystics have also predicted and for the same reason - the end of a cycle and the start of a new one.

December 21, 2012 (the winter solstice for the Northern Hemisphere) marks the last day of a 5,126 year *cycle.* That cycle began on August 13, 3114 B.C. (according to our modern Gregorian calendar) and ends on December 21, 2012 A.D. It's one of five smaller 5,126 year cycles that make up a larger 25,630 year cycle. Each of these 5,126 year cycles represents a Great World Age, which is reflective of a common way of thinking - a collective-mindset.

The cycle we're ending on December 21, 2012 represents a massive 5,126 journey into impurity. The Darkness of the last 5,126 years of human civilisation can verify this. This 5,126 year journey has shown the human race the sort of weak, foolish, serious and hateful thought-action-feeling that we *don't* want - otherwise how would we know and why would we seek?

December 22, 2012, starts a brand new 5,126 journey *back* to purity of thought, where the human race seeks powerful, wise, joyful and loving thought-action-feeling. That's what *we will all* want. That's what *we will all* seek. Can you feel that urge? Are you fed up with negativity of the modern world? Do you feel a calling? Do you feel the change? That's the urge within the collective-mindset, already active.

The new 5,126 journey *back* to purity of thought is merely one section of a much larger 25,630 year cycle of en-Light-enment. On a human level, we spend one fifth of *our* lives (1-21 years) developing an EGO and then the other four-fifths (22-105) resolving that EGO to achieve en-Light-enment. The Galactic Cycle does the same. The Galactic Cycle just spent the last 5,126 years developing a World-Age of EGO (this represents the one fifth) and is now starting a 20,504 year Great World-Age of en-Light-enment (this represent the four-fifths, 4 x 5,126 = 20,504).

The Great World-Age of en-Light-enment marks the point where the human race *begins* to re-member what it *collectively* forgot - the Truth. The human race will begin to re-member it's potential, it will re-member joy, it will re-member love and it will re-member what it is to *be* GODlike. Eventually (maybe even in this century), the human race will also re-member that they are Gods-in-human-form and that they are all connected as *one.*

It may be the 'End of the World' but it's the 'End of a World of Darkness'.

WE'RE STILL HERE

People have been predicting the end of the world for thousands of years and yet *we're still here*. When people ask me about my view on the end of the world, I always answer with, 'Which do you think is more probable?':

> **SCENARIO 1** - A bunch of Gods with infinite power and wisdom suddenly deciding to wipe out the human race and maybe even the planet they co-created; effectively ceasing the dualistic journey that was beautifully assisting billions of Souls to re-member higher and higher levels of potential in order to expand the entire Cosmos for all beings; *and* helping GOD have a disconnected and impartial experience of Her own Divine Magnificence?
>
> **SCENARIO 2** - A bunch of humans being influenced by fear and predicting a whole lot of gloom and doom?

Of course it's scenario 2. The strongest negative element of the EGO is fear. Fear is an all-pervasive, suffocating, bad habit of reacting to life with negative thought. It's the base of the EGO and it activates and feeds and strengthens every other bad habit of negative thought. It creates a fear of fear itself, a fear of lack, a fear of failure, a fear of success, a fear of speaking, a fear of conflict, a fear of knowing and a fear of not-knowing. It makes humans *panic* about death and it's the fundamental source of most conspiracy theories and end-of-days predictions.

The Cosmos was constructed from powerful and wise potential. It's only intent is the eventual re-membering by all beings of this powerful and wise potential so that each being can feel joy and share that joy as love. Sure, the Cosmos recognises that extreme-duality is an essential part of the en-Light-enment process but not *that* extreme that it wipes out the population of the planet. That's just plain stupid because there is no *challenge* in that. Death is the *easy* way out. Life itself *is* the challenge. That immediately rules out the possibility of the human race being wiped out by an End of Days catastrophe - in 2012 or later. It's onward and upward via the purposeful and necessary challenges of *life*, not 'Oh crap, let's start again with a new set of humans'. The same thing applies to wiping out Mother Earth. What good would that do?

We're no good to the Cosmos dead! As far as the expansion of consciousness ... we're it. Everything else is pure. We're the only beings capable of creating impurity, because we're the only beings capable of thinking negatively, because we're the only beings who were brave enough to forget the Truth and incarnate as humans *being* impure. We create the impurity because it produces extreme-duality. Our reactions and responses to that extreme-duality are what challenge us to re-member potential. That re-membered potential collectively expands the Cosmos, *for all other beings.* In other words, we're spiritual warriors, we're special, we're needed here, and we need to be *alive*, living on a planet, to do that! This means:

The world is not going to end. We are only ever going to evolve into higher and higher levels of power, wisdom, joy and love and we'll continue to incarnate here, as humans, on Earth, until we do. That's all that has ever happened and that's all that will ever happen. The future of Mother Earth is guaranteed and the future of humanity is guaranteed. Everything is running perfectly, because it's run on a perfect system engineered by GOD and facilitated by Gods.

In short, the world's not going to end. That's ridiculous! If anything, there is more chance of us living *longer,* on a perfect, beautiful, planet Earth, in an ever increasing state of joy and love because that's what's required for *further* expansion of our potential.

If I'm wrong, before or after 2012, you're welcome to crawl out from underneath the rubble, bandage your life-threatening wounds, find some sort of power source, search for a working mobile phone and then ring me, hoping that the mobile phone network is still intact, that I too have a working mobile phone with me, battery charged *and* that I have it switched on during a cataclysmic disaster and tell me I was wrong! In other words, I think you will have other, more important things to worry about.

THE GAME HAS JUST BEGUN

Humans are obsessed with their own specialness. But as far as potential evolution is concerned the 'Game of Life' for humans has only just begun. In the last 5,126 years, we've only re-membered a *tiny tiny tiny* amount of our GODlike potential. We're only now *just* beginning our 20,504 year journey of collective en-Light-enment.

Heck, we're still fascinated with Internet porn. Did you know, at the end of 2011, the word 'sex' had a billion global *monthly* searches on the Internet (probably more!). To put that into perspective, if you banged a hammer on a desk every time a human searched for the word 'sex' on the Internet in 2011, it would create a sound with a bang-bang-bang frequency, thirty-eight times faster than a Kalashnikov AK47 assault rifle on rapid fire! That's three hundred and eighty 'bangs' a second. Instead of a 'rat a tat tat' machine gun sound, it's a 'burrr' vibration with a frequency very close to the middle keys on a piano keyboard.

We will reincarnate here for thousands of years until the human race eventually glows with Light - and I mean that literally. We'll eventually re-member so much potential that we'll bypass the Jesus-like state (physical mastery of power, wisdom and joy). Our bodies will then vibrate at a frequency outside the current levels of human perception and disappear from view. We'll then glow with Light, as Gods again, either in the astral-world or the spirit-world. When all of us do that, collectively, it will signal the end of our human journey on this planet.

Then a new multi-billion year 'Game of Life' will begin with a new race of ape-like humans, who will eventually evolve into human-beings with no idea a race of humans-beings lived there, before them. *We'll* still be there, guiding the new ape-like humans from beyond their perception, just like the humans before us *do now,* as Gods in the spirit-world.

In that respect, our 'Game of Life' has only just *begun*. To put human evolution into some sort of perspective, just imagine the entire 4.6 billion year evolution of our planet was crammed into a twenty-four hour period. Imagine if 00:00 hours represents the formation of the Earth 4.6 billion years ago and 24:00 hours (midnight) represents present day.

On this imaginary time-line scale, the Earth formed at 00:00 hours. The Earth didn't even cool down enough to develop a crust until 02:36 hours (2:36 a.m.). A few very simple signs of life appeared on that crust at 04:10 hours (4:10 a.m.) and evolution *began* with sexual reproduction at 17:44 hours (that's 5:44 p.m. - seventeen hours and forty-four minutes *into* the twenty-four hour timeline). Simple animals appeared at about 20:52 hours (8:52 p.m.) followed by some 'big arse' insects at 21:01 hours (9:01 p.m.). Land plants appeared at 21:31 hours (9:31 p.m.) and frogs and reptiles at 22:26 hours (10:26 p.m.).

Dinosaurs popped in at about 22:49 hours (10:49 p.m.) to eat everything and stomp around a lot. Mammals showed up at 22:52 hours (10:52 p.m.), to help feed the bigger dinosaurs, birds at 23:13 hours (11:13 p.m.) as snacks, and by 23:19 hours (11:39 p.m.) a ten kilometre wide asteroid with an estimated explosive force of a billion megatons of TNT, supposedly wiped out the dinosaurs (probably as karma for eating all the furry mammals and birds).

At 23:59:22 hours (11:59:22 p.m.) our common ape-like ancestor appeared wondering where all the big lizards went. We didn't even develop into the anatomically modern human in Africa until 11:59:56 p.m. On this twenty-four *hour* imaginary timeline, that's four seconds before the end of the twenty-four hour period! Four seconds to go and we *just* appeared as Africans!

But what really makes you gasp in horror is that the Industrial Revolution and the subsequent pollution, extinction and depletion of resources needed to fuel that 'revolution', happened at 23:59:59.9967 p.m. That means that on this imaginary twenty-four time-line, we humans have managed to almost completely destroy the eco-system of this planet, in the last three one-thousandths of a second! Humbling isn't it, especially considering that three one-thousandths of a second is one hundred times faster than the blink of an eye?!

SUMMARY

- Fifteen thousand babies are being born every hour, adding a lot of mind to the planet.
- All minds and the potentials within these minds are connected, as *one* collective-mindset.
- The potential energy within the collective-mindset accumulates and strengthens to a level that eventually and collectively influences the brains of *all people,* to think in a similar way - regardless of whether or not they even have potential within them.
- The net effect is a *parabolic increase* in a certain type of thinking.
- This is the 100th Monkey Syndrome and the tipping point is December 21, 2012.
- From that point on there will be a parabolic increase in power, wisdom and joy across the globe.
- This is The Potential Age.

- It's a great time to be alive.
- The change will be individual and collective.
- Collectively the human race will awaken.
- Individually, you can embrace that potential now and soar.
- If you use that potential to do what you love because you know what you love doing, in order to serve others in a state of joy, then you will feel levels of joy and love beyond your imagination.
- The change will be slow but parabolic just like the collective-mindset of the environment.
- It might take 21-42 years to accomplish a *noticeable* integration of this new potential – *collectively.*
- Within the first 21-42 years of The Potential Age, people will reach the Jesus-Like state.
- This marks the 'Second Coming' or the 'Return of the Light'.
- It's more likely the children being born today will have the potential evolution and time to achieve a Jesus-like state by age 33 years.
- The rest of us may be able to achieve this state if we fully embrace the new collective potentials of The Potential Age.
- It's more likely we will have Jesus-like moments.
- Don't jump the queue in Church. A priest might yell at you.
- A lot can happen in 29-35 years.
- Modern communication technology was Divinely inspired to help spread the new ways of thinking of The Potential-Age.

- The World Wide Web will become globally collective at the exact same rate the human mind becomes potentially-collective.
- The only side effect of The Potential Age is an increase in dis-ease in those who resist the new potentials. That resistance may create a disease, which then becomes a necessary-challenge drastic enough to shift them out of the EGO that created the original resistance.
- The Internet and modern communication will help spread the powerful and wise message of The Potential Age across the globe.
- The Mayan calendar predicts the end of a cycle - not the end of the world.
- December 21, 2012 marks the last day of a 5,126 year Great World Age of Darkness and signals the start of a new 5,126 year Great World Age of en-Light-enment.
- The world is not going to end.
- The challenge is life, not death.
- On an imaginary twenty-four hour timeline, the human race has almost completely destroyed the eco-system of this planet, in the last three one-thousandths of a second.

DEATH

You'd better en-joy life as much as you can because it ends abruptly with death - guaranteed. Out of respect for grieving relatives and for people who have been deeply affected by death, I must point out that I don't handle the topic of death very delicately. I consider death to be the most natural, the most beautiful and the most amazing of all experiences. I have no doubt of that.

Most conversations between people are based on the four major elements of life - yourself, other people, the world and things. People love to talk about themselves, their relationships, the latest gossip or what new thing they bought. For example, if you start a rumour about how you had sex with a stranger in the car park while shopping for a new iPhone™, it will get people's complete and undivided attention because it contains all four major elements of life.

With that in mind, I find it really interesting that no one talks about death. And I mean, *really talks* about *their own* death, in-depth and with the same sort of enthusiasm as, 'Hey, guess what? I bonked a stranger in the shopping centre car-park and filmed it on my iPhone™'. The reason I find this interesting is because death must *surely* be *number one* on the list of 'Things That Most Affect You, People, Your World and Things'.

It's not like we don't *know* about death. We're surrounded by it. Every single living thing on this planet dies, including us. We're glued to the TV watching crime shows about murder. Just about every major film has a death scene or a strong connection to death. We're fascinated by ghosts, graveyards and the history of death. There's no greater gossip than, 'Did you see the tragedy on The News? Twenty-five people died!', or 'Did you know the old guy next door died last week?'

Yet we don't talk about our own *imminent* death. Why? Because death frightens us. We don't know what happens *during* death. We don't know if death hurts, if it's unpleasant or if it's shocking. We don't know what happens *after* death. We don't know if there is an afterlife, if we go to 'Hell' or 'Heaven' or some sort of teenage Purgatory for breaking any one of the myriad of rules we were supposed to have respected in our youth.

Consequently, the subject of death is taboo. We'd rather ignore it, hope it goes away and bothers the old guy next door. We don't want to discuss the idea that life has an expiry date. We *just don't like* talking about death. So, let's talk about death.

We all die. You will die. That's a fact and there's no escaping death. But death is not the end. Death is simply a transition, via reincarnation, into a new life. We even symbolically plant ourselves in a coffin in the ground like a seed and then cover ourselves in dirt, so we can grow 'somewhere else'!

Does it hurt when you die? I don't know. I've heard that the conscious mind leaves the body just prior to a horrific death to save you the pain. It makes sense because at *that* moment in your life there is no point learning anything new, so why would your Soul put you through that necessary-challenge? That makes sense to me. My heart tells me that death feels *absolutely amazing*.

What I do know is that the death process is effortless, painless, peaceful and *fast* (one nonillionth of a second). It's the most natural, the most beautiful and the most amazing of all experiences. I know ... I went there in November 2003.

THE COSMOS WON'T KILL YOU

The Cosmos is not out to kill you. It plays no part in your death. Neither do the Gods and especially not GOD. They're 'cheering for you' the entire time, calling you towards the purity of *life*, not death. They want you to stay alive and for *as long as possible*.

You choose your own self-inflicted death, via the EGO. The EGO urges negative thought, which creates negative e-motion, which smothers your cells in toxic low-frequency vibration. That toxicity ages and destroys your cells and being a conglomerate of cells that ages and destroys you - slowly and gracefully.

From the age of twenty-one onwards (the year your EGO becomes fully formed), your EGO ages you, so that you slowly and gracefully age, mature, get old and die. That's actually one of the primary roles of the EGO - to ensure your death makes way for new life and to ensure we don't saturate the planet with too many humans. The more EGO you have the faster you age. The more joyful you are, the slower you age. The choice is yours. If you achieve en-Light-enment, you may even halt or *reverse* the aging process.

Either way, you're *supposed* to get old and die peacefully, just like every other animal on the planet. Old animals know when they are about to die. They just find a nice cool place, lie down and die peacefully. We should do the same, although not on the grass in front of your home, but maybe in a nice coffin with family present.

Your EGO can kill you faster, if you choose. Your EGO can kill you *directly,* via your own recklessness, apathy or ignorance. It can kill you *externally,* via your co-creation of a world that attracts violence, accidents or disaster into your life. It can kill you *internally,* via the creation of dis-ease. It can kill you *habitually* with cigarettes, drugs, over-eating and alcohol. If you've got a really big EGO, it can kill you *spectacularly* from life-threatening wounds sustained from crashing a triple somersaulting snow plough. Once again - your choice.

THE DREAM WORLD

To understand what happens when you die you first have to understand what happens when you dream. The conscious mind is your now-focus. Wherever your conscious mind *is,* is where *you* think you are, *now.* When you're awake, your conscious mind is focussed in the physical-world and so you think you're in the physical-world.

Normally, your mindset pulses between the physical-world and the astral-world, to accommodate the powerful potential of the conscious mind and wise potential of the subconscious mind. When you sleep, the pulse shifts so that your conscious mind leaves the physical-world and becomes focussed in the astral-world. Your pulse shifts up the scale to pulse back-and-forth from lower-astral-world to upper-astral-world, instead of pulsing back-and-forth from the physical-world to the astral-world.

When your conscious mind enters the astral-world, the *you that you think is you* (or *you* for short) is no longer now-focussed within the physical-world but is now-focussed within the astral-world. The astral-world is the subconscious mind so when *you* become focussed in the subconscious mind, *you* become focussed within the past *memories* of life.

Memories are simply accurate records of what was real. When you dream, *you* stand 'knee-deep' in these memories and interact with these memories, as if they were real. This is why dreams can seem *so real* - the memories you are standing in *are* real.

In the real-world, what we think about co-creates events in our lives that manifest in a 'normal way', congruent with space and time. In other words, it takes time and involves real-world events. For example, riding a horse will only happen if you desire to do that, take action, wait for free time on the weekend to do so, drive to the horse park, sit on a horse and ride it.

The astral-world is a non-physical-world and so is not limited by the physical-world constraints of space-time. It operates on the *memory* of space-time. A memory of space-time is an unreal perspective of space-time. It's weird. You can bypass the normal effects of that space-time. You can move through space and move forward or back in time, at a whim. For example, if you want to ride a horse in your dream, seconds later you could be riding a horse. There is no need to wait or drive there or queue up or get on the horse, etc. It can happen there and then.

What you create in your dreams depends on what you think about in the dream. What you think about in that dream is triggered by what you were thinking about before you went to sleep. For example if you go to bed stressing about work you'll probably have a dream about being stressed.

That feeling of stress can activate any subconscious memory that has been previously associated with stress. That's any stressful fear, fantasy, person, event or vision stored in your subconscious mind, from your adult life *and* your childhood life. That's where dreams can get really weird, really fast. One minute you're dreaming about being stressed at work, the next minute you're riding a horse through your office, side-saddle, naked, breast-feeding a duck!

You stop dreaming when you're woken by disturbances in the real world. For example, whilst you're sleeping, your bedroom door might slam shut in the breeze and that noise will awaken you, ceasing the dream. But the speed at which your conscious mind perceives the outside world and then stores that information in your subconscious mind as memory, is staggering.

For example, a door slamming in the real world is only a millisecond event. That event gets stored in your subconscious mind as sound-of-the-door-slamming, at such a staggering nonillionth of a second speed, that it can feed your dreams, *faster* than the millisecond duration of the actual banging-door sound. The dream you're in, might shift to a new scene, triggered by the memory of the sound-of-the-door-slamming. You might then dream of a door slamming, in your dream. That dream might then continue for what seems like hours, in 'dream land'. But when you awaken in the real world, the door is *still slamming*! Ponder that!

The moment your dream ends is the moment your conscious mind now-focus returns to the much denser physical-world. This sudden lowering of vibration often creates the sensation of falling (a sudden lowering). This often creates dreams of falling just as you are waking or a sudden jerk of the arms or legs when you're in bed, especially if you're dreaming about falling.

DREAM INTERPRETATION

Dreams allow the conscious mind to leave the physical body while you sleep. Not having the mindset change states from a dense physical-world-conscious mind to lighter astral-subconscious mind, nonillions of times a second, gives the physical body a much needed electromagnetic rest.

That doesn't mean that when you sleep your body disappears into the astral-world. The form, function and phenomenon of your human-body is always being maintained by vibrational-reality created by the pulse of your contemplation-mindset.

As far as dream interpretation is concerned, my personal opinion is that most dreams reflect your thought patterns rather than any great spiritual revelations. You can get much more en-Light-ening revelations, right here in the physical-world by using your own powerful and wise thought-action to overcome, facilitate or master a challenge - often far more enlightening than unravelling why you were breast-feeding a duck.

This is not to say that dreams aren't important. On the contrary, standing 'knee deep' in your memories gives you a 'real-life' preview of your habits of thought. Your dreams replay back to you, in a safe environment, the wonderland of OPTA or the nightmare of your EGO.

WHAT HAPPENS WHEN YOU DIE?

The really big question is: 'What happens when you die?' This I can answer. November 2003 was also a *living* near-death-experience. For a nonillionth of a second I went to a place that dead people go but I was lucky enough to come back.

When you die, your pulse shifts so that your conscious mind leaves the physical-world and becomes focussed in the astral-world, the same way it does when you dream, except it stays there. You become consciously aware *within* the subconscious mind as if you were dreaming - except in this case you're not going to wake up because your conscious mind is not going back to the physical-world. Just like dreaming, you would be consciously aware *within* the subconscious mind, and 'standing knee deep' in your life's memories.

As your conscious mind shifts out of the physical-world and into the astral-world your vibrational frequency rises, rapidly. You become Lighter as your frequency rises, similar to how 'light' you feel when you are happy versus how 'heavy' you feel when you're depressed. This combined with the sudden exit out of the physical-world, creates the sensation of floating, flying or ascending (hence the near-death-experiences of people floating above their 'dead' body).

What you experience in the astral-world after death will depend on *what you believe.* Whatever you think about in the astral-world manifests instantly as the 'real world', based on your past memories. If you believe when you die you go to Heaven then you will '*dream*' of Heaven, fuelled by your past memories of all-things-good and your imagination of Heaven. If you believe god is an angry god and will punish you in the fires of Hell, then you will experience *that* crusty-skin crackling 'nightmare', fuelled by your past memories of all-things-bad and every memory of feeling pain from a burn.

I choose to believe that Heaven is a white room, filled with all the people I love who have died, who walk toward me carrying golden trays filled with freshly cooked Western Australian crayfish, cheese, and every gadget I have ever dreamed of owning - mainly because that's what I would like to experience in the astral-world after death.

YOU'LL FIND PEACE

The good news is, as with any dream, you will eventually awaken and *you* will realise where you are and why you are there. You are there to review your life in order to find peace. The life-review is the second stage of the death journey. I also believe this to be the Catholic idea of Purgatory, the 'place you go before Heaven'.

During your life-review, your life will flash before your eyes, including the moment you died and ascended out of your body. This is more likely the way people would experience the sensation of floating above their dead bodies. They're re-witnessing those last moments as the *memory* of those last moments is *now* being replayed before their eyes.

The reason why the memory files flash before your eyes from birth to death, rather than just randomly, is because your conscious mind's rising vibration chooses memory files of a higher and higher potential, which is usually in order of evolution and age.

The astral-world is the dream-world and so when you review your life's memories in the astral-world, you'll be able to interact with these memories, including the people and events within them, like you do when you're dreaming. Events will be replayed, you'll be able to ask people questions and you'll review as many memories as is required, to help you understand how the flow-on effect of your actions helped *millions* of people re-member potential and become more en-Light-ened.

Going back to the example of, 'I Washed My Hands and Brought Down the Stock Market!', you will witness and interact with the people whose lives were changed by such an event. For example, the man in the motel who was awakened by the squealing water running through the pipes will tell you how his bad day led to a whole series of events that made him finally confront his boss about working conditions.

He never 'rocked the boat' prior to that event because he was afraid of standing up for himself. On that day, his anger overpowered his depression, giving him the courage to say *what he needed to say*. He got angry enough to blurt out his concerns. His anger was enough to get his boss's full attention. The boss listened carefully to 'the nice guy that never said much' and then decided to give this guy better working conditions. That man learned that expressing himself not only made him feel better but it was a very commanding way to create change.

The stock-market trader, who forgot the important trade, will tell you how he spent three days in hospital, lying on his back, with nothing else to do but *contemplate life*. He was always too busy to do so. He never stopped long enough to re-evaluate things. His contemplation gave him a chance to really think about what's *really important in life*. He consequently decided to spend less time at work, to de-stress, and spend more time with his family.

Even total strangers were affected. One man will tell you how that sudden stock market crash knocked thirty-five per cent off the value of his shares-based superannuation. This 'forced' him, at the age of sixty-eight, to find part-time work. Yet, unknowingly, the security potential he had to embrace in order to urge him back to work flooded his lower body with positive e-motion, reversing a potentially lethal cancer growing in the base of his spine. That man cheated death and got to spend years more time with his grandson. He mentored his grandson as a wise elder, steering his grandson away from an imminent self-destructive path of drug induced, depressed laziness.

You will have the same experience. You will see how *all* your actions contributed to a massive collective flow-on effect that *eventually* affected every human on the planet - we're all connected in some way. You will witness how your actions, good or bad, led to events that *challenged* these people, good or bad. You will see how those challenges helped these people to re-member potential. You will see how that led to joy then joyful service and then love.

The process will last as long as required for you to find peace. Peace will come from understanding *why* things in your life, happened the way they did. It's the reward you receive after exiting the chaos of extreme-duality. With a full and compassionate understanding, you will be able to let go of all your regrets, guilt, worry and stress. You will feel full and complete forgiveness for all. Most importantly you will feel *at peace with yourself.*

Your life-review pacifies your conscious vibration and prepares you for the next stage of death, which you cannot enter unless you are in a state of peace.

THE GATEWAY OF DEATH

The third stage of the death journey is to leave the astral-world and enter the Cosmic-pond. In death, the Cosmic-pond appears as an infinite dark tranquillity, a place of incredible peace and stillness, (hence near-death-experiences where people describe a great peace within infinite darkness). The Cosmic-pond acts like a gateway of death, otherwise known as the Gates of Heaven, or 'The Veil' and you cannot enter this place of peace unless you are *at peace* with yourself (hence the life-review).

This is the last chance. Once you leave this place and cross the Cosmic-pond into the Light of the spirit-world, there is no going back. During this last chance, you get to choose whether you wish to continue into the spirit-world (death) or return to the physical-world (life). This choice is often guided by your Soul-Group who will ask you to choose between life and death.

The Soul-Group will appear to you in whatever format you believe such as Saint Peter (the keeper of the 'keys to the kingdom'), Angels, or deceased relatives. There are many near-death-experiences that describe a dead relative (usually a deceased parent or grandparent) offering soothing advice from the Light (spirit-world) such as, 'Yeah come over to the Light, it's amazing', or 'It's not your time to die yet, go back to life. You'll be OK'. These same near-death-experiences would often describe how difficult a decision that was to make because they could feel the love coming from the Light and wanted to 'go there instead'.

The Christian doctrine about being denied entrance into Heaven and then being cast down into Hell for your sins, refers to what happens if you refuse to acknowledge you're dead. If *you* cannot accept your death, then you will never reach the Cosmic-pond, because you are not at peace. In that regard, the Cosmic-pond does act as a gateway because without peace, you get 'refused' entrance into Heaven. You will remain in the astral-world until you 'wake up', realise where you are, why you are there, and then accept your life review. If you don't wake up and you believe in Hell, then you may well create Hell for 'all eternity' and be punished for your sins (you will dream your own negativity until you wake up).

Helping a dead person 'go to the Light', refers to helping a dead person 'awaken' in the astral-world, realise they are dead, agree to a life-review, find peace, go to the Cosmic-pond, and then make a choice to either cross over the pond and 'go to the Light' of the spirit-world, or choose to continue their incarnation by coming back *into* their bodies, wake up in the physical-world, and continue living.

Prayers for the dying, spiritual work and the thoughts of grieving relatives all get added to the collective-mindset, which is *in* the astral-world. The dead person is 'standing knee deep' in these messages and this can often help them to awaken from the nightmare of 'Hell' and 'go to the Light'. Personally, I believe once you die, there is very little chance of getting lost on the journey because the entire death process is guided by the Soul. I believe your Soul will be there *with you* the entire time, guiding you, comforting you and loving you towards the Cosmic-pond.

Either way, once you reach a state of peace during your life-review and reach the Cosmic-pond, *you* decide what you want to do - cross over into the spirit-world to die or return to the physical-world to live.

Once you decide to cross the Cosmic-pond and enter the spirit-world, your conscious mind will pulse into the spirit-world. You enter into the Light, which from a place of darkness would look like a tunnel of light (hence near-death-experiences about tunnels of light). That's exactly what I saw in November 2003 - the room disappearing, a darkness and then the sensation of going up into a tunnel of light, etc. My conscious mind, for whatever reason, came with me on a *one* nonillionth of a second pulse ride into the spirit-world.

Of course, I came back so I didn't actually die! Phew! My case was a little different. I'm still not one hundred per cent sure of *where* I went in November 2003. I know I was surrounded by Light, of that I have no doubt but I'm not sure *where* I was witnessing that Light from. It could have been from the spirit-world, it could have been from the Cosmic-pond or it could have been from the astral-world.

This information has not been revealed to me as yet because I get the feeling it's not important for me to know, because it's not important for others to know. The real challenge lies within the physical-world and there's where our Souls would like us to focus *more* of our attention, rather than *less of it,* trying to access some higher spiritual plane.

What I do know is that I was deliberately shielded from seeing *all* of the Light. I only saw the very lowest portion of the Light and only the wisdom aspect of *that*. If I had experienced the power as well, then I would have felt Cosmic joy. If I had gone higher I would have felt Cosmic love. If I had felt either, even for a nonillionth of a second, I never would have come back! It would have *felt* far too amazing for me to resist, I would have entered the Light and probably died.

When *you* die, you will get to experience *full* Cosmic joy and love, for as long as you wish and it will be mind-blowing.

GHOSTS AND SPIRITS

As soon as you leave the Cosmic-pond and cross over into the spirit-world, your subconscious mind separates from your conscious mind. The memories of human-thought stored in your subconscious mind are too impure to enter the spirit-world. Your subconscious memory is dropped off in the lower realms of the astral-world. It's at this moment your minds *disconnect*. Your pulse stops, your human-pulse stops (the heart), the signals in the brain stop and you are officially dead.

Your subconscious mind lies dormant in the astral-world. It's dormant because it no longer has a conscious mind and so can no longer pulse as a mindset. A non-pulsing mindset is a non-conscious being. Without its conscious mind-partner, it's just the memory of the person. I call this non-conscious, non-pulsing, disconnected chunk of subconscious mind … a ghost.

Your ghost lies dormant in the astral-world, until it's activated by a 'spare' conscious mind. In an environment of grief, it's often the grieving relative's conscious mind that activates a ghost back into life!

When a grieving relative thinks about a dead person, grieves over them or reminisces about them, their conscious mind searches for the memory of that person. That searching attaches the conscious mind of the grieving relative to the subconscious mind of the ghost. This allows the ghost to become a pulsing mindset again. A pulsing mindset is a conscious being, except in this case the consciousness is coming from the grieving relative, so it's not really alive just more discernibly active.

This is what's meant by 'keeping the memory of the dead person alive' or when a ghost 'attaches itself' to a human, when the possession is actually the human attaching itself to the ghost. The same applies to 'letting go of the ghost so it can go over to the other side - over to the Light'. It's the grieving relative that lets the ghost go, when *they* stop thinking about it. Really the ghost doesn't need to 'go anywhere' since its place is the astral-world.

This ghost is not a conscious being either, because its 'conscious mind' is not its own. That belongs to the grieving relative. From the ghost's perspective it would seem like it has a memory of *who it is* but confused about *where it is* and *why it's there.* Imagine how you'd feel if you remembered your past but didn't know where you were *now,* or *why* you were there. You would also feel lost and confused. That's why ghosts are often referred to as 'lost souls'.

The ghost still exhibits the character traits of the dead person because it's a chunk of subconscious memory - the dead person's memory. How it behaves will depend on what it's 'thinking now' and what it's 'thinking now' is determined by what the grieving relative is *thinking now.* The ghost is sharing the grieving relative's conscious mind, so it will be what the grieving relative is being. It could be happy or it could be sad. It could be loving or it could be angry.

PSYCHICS AND MEDIUMS

The entire point of having a ghost is to give the grieving relative a phenomenon they *can* experience in *the real world.* The *real* dead person, the conscious mind part of the deceased, has long gone. He/she has entered the spirit-world, reunited with their Soul and is now having a great time surrounded by Cosmic joy and love. They have become a Soul again; a God; they have gone home and they are OK.

Grieving relatives cannot have direct contact with a Soul. It's too scrazoompity frrzzzerktaty and would turn a grieving relative into a fried, bloody mist. That just creates more grieving relatives and kind of defeats the purpose. Therefore grieving relatives cannot interact with the real dead person.

Humans *can* interact with ghosts. That's why we have them. The conscious mind of the grieving relative connects with the ghost and reactivates that ghost. The pulse of the ghosts then creates real electromagnetic phenomena. The grieving relative can then interact with this phenomenon, in *their* world.

For example, grieving relatives can often feel e-motions created by the ghost. They can often picture them in their mind or even talk to them via their thoughts. They can often smell familiar scents from the ghost, such as tobacco or perfume (scent is just an electromagnetic vibration). They can sometimes catch a glimpse out of the corner of their eye. There are many cases of ghosts creating miracles, manifesting familiar objects or leaving a body-weighted impression in the sheets as the ghost sits on the bed.

Most importantly, grieving relatives can interact with the ghost in their dreams (ghosts live in the astral-world) as if the ghost is a real person. They can then express how much they miss them, they can hug them, they can ask questions or get advice, or they can even get angry at them.

Ghostly interaction was originally designed by the Gods to help people grieve. Grief sucks. There is no *cure* for grief. Grief is a very real phenomena caused by the electromagnetic ties between two people being severed by death. The conscious mind has been removed from the collective-mindset and so even strangers can feel this loss. You can heal grief (lessen the pain over time) but you cannot fix it. It's always there, it hurts, it sucks and it's just part of the death process! Any tool that can help *ease* that kind of pain is valid.

If a grieving relative tries to block grief by masking it, denying it or deflecting it, it will create dis-ease. Grief is a very low vibrational emotion. It needs to be expressed so its vibration ceases and the e-motion disappears from the body. If the e-motion of grief is resisted, it just keeps on repeating itself, flooding the body with more and more toxic negative e-motion, which creates dis-ease.

The fastest way to break this cycle of grief is to cry. Crying, along with anger, is a physiological release of negative energy. But people, especially men, don't like to cry. They resist it with gritted teeth, pursed lips, or sobbing opposition to it. It's almost like people do their best to resist crying until something 'pushes them over the edge' and they burst into tears. Then they let go and let it flow. The feelings and memories reactivated from ghostly-interaction is often enough to 'push a grieving relative over the edge' and make them cry, mourn, hurt or get angry. It helps them to grieve.

In a time of grief, a grieving relative often has trouble connecting with the higher vibrations of the astral-world making it difficult (but not impossible) to interact with the ghost. This can account for the shock or indifference of grief. The grieving relative can sometimes be in such a state of grief that their low vibrational energy-state prevents them from connecting with the much higher vibrational-frequencies of the astral-world - where the ghost lives. Without astral-world interaction they go numb and it often stops them from grieving.

A psychically sensitive psychic-medium often exhibits a vibration high enough to tap into the astral-world. They can then connect with the ghost and facilitate communication between the grieving relative and the ghost. It helps the grieving relative 'talk directly' to their loved one and express what they need to express. It offers 'proof' that there is an afterlife and helps reassure them that their loved one is OK and has moved onto a 'better place'. Good psychic-mediums often tend to be experienced with grief-counselling and can help grieving relatives let go of the guilt and anger associated with death. In a time of grief, this is very comforting.

How does it work? The grieving relative and the ghost are sharing the same conscious mind. Whatever the grieving relative thinks and feels, the ghost 'thinks and feels'. A psychically sensitive medium can tune into the 'thoughts and feelings' of the ghost and then relay that psychic communication to the grieving relative. The message given then *exactly* matches what the grieving relative is thinking and feeling and in a format that matches the ghost's personality. That message then appears to be 'psychic'.

Often, the medium is picking up on the thoughts and emotions of the grieving relative sitting in front of them, rather than the ghost, but they're one and the same thing.

For example, a medium will often pass on the message from the ghost that says 'to stop feeling guilty for what they have done, or feeling sad, or confused' if there is a lot of guilt or grief *being felt by the grieving relatives*. A medium will often pass on the message from the ghost that says 'stop being frustrated and angry' if the grieving relatives *have not dealt with grief-associated-anger*.

Mediums can also pick up on happier messages such as a ghost as being 'very loving and forgiving' if there is a lot of love being felt by the grieving relatives, or wise and enlightened answers if the ghost is called upon to answer questions expressed by a group of spiritual people.

This may also help explain why psychics and mediums might say that a ghost seems sad, angry or worried. The real dead person has returned to the spirit-world and reunited with their Soul, as *one*. Souls NEVER display negativity. They can't since they're 99.999999999 … % pure. Only ghosts will display negativity and human traits of behaviour.

SOULFUL COMMUNICATION

The *really good* psychic-mediums *can* communicate on a Soul level. They can connect to your Soul and then communicate information to you that's 'out of this world'. In other words, they can interpret and communicate downloads, *for you*. The difference between Soulful communication from Souls in the spirit-world and psychic communication from ghosts in the astral-world, include the following:

- Soulful communication will NEVER display any sort of negativity, especially anger, sadness, guilt, confusion or impatience – EVER.
- Soulful communication will ALWAYS feel amazing - an overwhelming feeling of love and joy that's incredibly powerful yet gentle, profoundly wise yet humble, and always patient, compassionate and grateful.
- Soulful communication will NEVER tell you what to do but often inspires you to ask the questions for yourself, because it values the challenge of you making decisions for yourself.
- Soulful communication will NEVER interfere with the life of another person, or give you personal information

about another person, because it values the privacy and free will of others.

- Soulful communication will NEVER offer advice that cheats the system or cuts corners, because it values the challenge of purpose.
- Soulful communication will NEVER tell you when you're going to die, the lotto numbers, who is your true love, if your marriage is OK, or what your future holds because it values the challenge of you finding out for yourself.
- Anything else is psychic communication from the astral-world, gleaned from the re-activation of non-conscious, disconnected, chunks of subconscious mind memory files that's often distinctly human.

If a good medium can communicate with your Soul they can also communicate with the Souls of people who have died, since they are all connected as *one* in the spirit-world. This represents the highest form of mediumship where the medium connects with the Soul and the Soul communicates directly to the medium but via the personality of the ghost. In other words, the ghost has been reactivated via the conscious mind of the Soul rather than the conscious mind of the grieving relative (or the medium). The net result is Soulful communication that displays the character of the ghost.

The Soul is driving the ghost and so the communication might resemble the drunken uncle or the cheeky aunt but yet display incredible power, profound wisdom and loving joy (compassion, patience, gratitude). It will be GODlike, yet human-ish.

But here again lies the bitter-sweet pickle. The communication between medium and Soul is never perfect because it has to be interpreted by the 'humanness' of the medium. They can only offer you Soulful communication at a level of accuracy that's directionally proportional to how joyful *they* feel and you joyful *you* feel, at the time of the reading.

In a state of grief, that state of joy required to achieve this level of communication is virtually impossible, both for the grieving relative and the psychic-medium seated in front of them. The e-motional environment of grief will smother both parties in low vibrational energy, lowering their vibration. This often means the psychic-medium will only be able to connect the ghost on an astral-world level and offer you ghostly communication. A ghost cannot *really* give you anything more than what they did when they were alive. Although that's often very comforting, it's not mind blowing.

If you want Soulful communication, my advice is NOT to go see a psychic medium in the initial stages of grief (or any strong negative e-motion). Energetically speaking, you're too 'heavy' and the low-vibration will pull the medium down into the astral-world. Wait a while, let some of the grief subside and then go see a psychic-medium when you're on a better, higher state of joy (or as happy as you *can* be).

EVIL SPIRITS

What is an evil spirit? Remember that a ghost lives in the astral-world, which permeates the physical-world. The only reason we can't see them is because they're vibrating at a different polarity than the physical-world and on a different plane of space-time.

There are very few feelings as intense as the feeling of grief and there is *immense* vibrational energy within the negative e-motions that make up the feeling of grief such as loss, regret, guilt, and anger. A ghost thinks and feels what the grieving relative thinks and feels. There can sometimes be so much amplitude within the e-motion of grief that it can induce equivalent energy levels in the ghost. So much so, that the vibrational energy of the ghost transcends dimensional separation and affects this dimension.

Remembering that vibrational energy is a real world force and so if that energy is powerful enough to transcend dimensional separation then it's often powerful enough to rattle windows, make people feel ill or dizzy, make strange noises and go bump in the night, move furniture, drop the temperature of a room, manifest objects or even create ghostly apparitions. This may account for 'evil spirits'.

That is very rare since the astral-world is *on a different plane* of existence. It takes a huge amount of energy to transcend this barrier. It would be the equivalent of loud music vibrating the windows of a house, ten blocks away. Why do you think humans hardly ever see ghosts? Why do you think no human has ever taken a good clear photograph of a ghost?

The only times we do see a ghost is when the feelings surrounding the ghost's death are *extremely* negative and collective. For example, the ghosts of murder victims or children or other tragic deaths. Many people *thinking of the ghost* (the terrible e-motive circumstances surrounding the infamous death) collectively and consciously pulse the subconscious mind of the ghost with so much energy that it can transcend dimensional separation. It can manifest (just) in the physical-world as ghostly apparitions and sometimes even do ghostly things. That's keeping the memory of the dead person alive (literally).

Imagine a high frequency, low amplitude vibration such as the vibrations of a men's electric shaver, a note played on a violin or a person's voice. Now image a low frequency, high amplitude vibration such as a washing machine on an unbalanced spin cycle, a jack hammer, or an earth quake. Which vibration do you think is going to move stuff? The answer is the low frequency, high amplitude vibration.

It's far more likely that strange happenings and things that go bump in the night are created by the intense amplitude of low-frequency vibrations, created by negative human e-motion, rather than ghostly forces. Negative e-motion is loud and powerful enough to do stuff. How many times has your computer crashed when you're in a really bad mood? Have you ever walked into a room after an argument and felt the thick, heavy energy in the room? Why is it that just about every confirmed case of a poltergeist 'moving stuff' also had an angry hormonal teenager(s) staying in the same house?

When you put it all into perspective there is *really* nothing to fear. A ghost can create intense low-frequency vibration, but so can you! In fact, that's your job! You incarnated here as a human *being* impure to create negative e-motion. You're an expert at it. You can handle it. A ghost cannot hurt you any more than you can hurt yourself with your own negativity. You are in more 'danger' in the physical world from self-inflicted dis-ease, than the 'evil' of ghosts.

If you're still frightened by the idea of ghosts, spirits and the spirit-world then remember ... the *real YOU is a* spirit, which *lives* in the spirit-world. YOU are one of the most powerful spiritual beings in the Cosmos and your Soul is like the Clint Eastwood of the spirit-world. Its Light wipes out *anything* negative. If you're ever worried about spirits or evil, simply ask your Soul to take care of the situation with its Light. Nothing 'evil' can survive YOUR Light.

In saying that, I would never suggest 'playing around' with astral-world negativity, especially séances, ouija boards or black magic. The astral-world is a collective-mindset. Tapping into astral-world forces means tapping into *massive* amounts of collective-negativity. That's never a good thing.

If these experiences are done in a state of fear or grief (as they often are) then the negativity of your fear makes you susceptible to collective-astral-world-negativity. You can easily handle negativity from your own mindset and maybe the mindsets of others around you but you are definitely not equipped to handle thousands of years of negativity, from billions of people, thinking millions of thoughts, that's now collectively stored within the astral-world. That can make you very sick or mentally ill.

My advice, forget about the astral-world, spirit-world, ghosts and all that other kind of non-worldly stuff. Instead, completely focus all your energy on being as powerful and wise as you can within the *physical-world*. That's why you're here, as a physical human-being, within a physical-world. Even your subconscious wisdom stored in the astral-world is there to help you make wise decisions, *now,* in the physical-world. Your Soul will take care of all that other non-worldly stuff for you.

YOU WILL NEVER KNOW

One thing is for certain, you will never know the time of your own death. Neither will anyone else. It will be presented to you as a surprise, by your Soul, as the greatest of all necessary-challenges.

Death is often used by your Soul-Group to great effect, especially as death affects so many people and at such a deep level. No death is accidental or wasted because death is a powerful way to remind people about the preciousness of life.

For example, two Soul's get together in the spirit-world. One Soul decides its human-being will die walking along the street and the other Soul decides its human-being will die driving a car. Perfect! Soul number one decides that its human-being will run over Soul number two's human-being, as Soul number one's human-being crashes into a tree and dies. Job done.

There is nothing like a tragic death to instantly awaken people to what's *really* important in life - peace, love, family, joy and life itself, because it comes as a direct result of the extreme-duality of carnage, grief, loss, sadness and death.

Life, peace, love, family and joy are all high outcomes that can only come from high levels of potential. To seek those things is to unwittingly seek potential. The Soul-Group knows we all have to die. The Soul-Group knows the shock of death is a very powerful way to make people search for the potential within life. The Soul-Group knows that if you knew the time and day of your own death, then you could alert both families prior to the event and that would dampen the shock value often required by humans to awaken to the joys of life.

NO NEED TO FEAR DEATH

There is no need to fear death. Ironically, only when you remove the fear of death will you start to live life because the fear of death is the major reason why people don't *challenge themselves*. However, I can understand why people fear death and so here is a simple tool to help ease the fear. Ask yourself, in any situation, 'What's the worst thing that can happen?' The answer is death.

From what I have just written you can see that death is going to be mind blowing (you wait until you read the last bits!). So what have you got to lose? The worst thing that can happen is actually one of the most natural, most beautiful and most amazing of all human experiences. You have nothing to lose and everything to gain by challenging yourself.

And not by triple-somersaulting a snow plough over a school bus, but by trying to be more secure, capable, expressive and knowing, to feel more joy, to serve others better, to feel more love. That's the *real* challenge and that's where you'll find en-Light-enment, peace, joy and love.

The word 'life' has an 'if' in the middle. Life is always a bit 'iffy', a bit uncertain, a bit risky, a bit of a challenge. Even when you find your purpose you never really know how it will turn out. So relax and have fun by doing what you love because you know what you love doing.

YOU REUNITE WITH YOUR SOUL

Back to the death-process. After you drop off your subconscious mind in the astral-world, the fourth stage of the death journey is to cross over the Cosmic-pond and enter the spirit-world. *You* would then be fully now-focussed within the spirit-world. *You* would see the Light again.

And because you entered the spirit-world without a subconscious mind, the urge of your conscious mind to reunite with another subconscious mind, would be overpowering. The *other* subconscious mind is your Soul, your *true* subconscious mind - the memory-of-potential. The last stage of death is when your conscious mind reunites, once again, with your true subconscious mind. At that moment you would be wholly (Holy) *one* again - a Holy Soul. That's when you feel Cosmic joy and love again.

Let's put that into perspective. Your Soul's memory-of-potential records every re-membered potential behind every thought *you* have ever made, from every lifetime, *infinitely*. In the spirit-world there is no space-time so your Soul can experience everything *you* have and will re-member, as the full memory-of-potential, from every lifetime, *all at once*. In other words, 99.999999999 ...% pure potential.

When *your* conscious mind reunites with your Soul's subconscious mind in the spirit-world after death, it reunites with the %99.999999999 ...% pure potential stored within your Soul's memory-of-potential. This means it can then consciously think with %99.999999999 ...% pure potential, producing thoughts of power and wisdom that would create pure desire, pure inspiration, pure joy and pure love - a joyful, blissful, loving ecstasy beyond human understanding. In other words, you will experience 'Heaven'. You will once again contemplate your own powerful and wise potential, as a God.

That's death! No matter how bad your 'sins' were as a human, no matter how many times you blamed, ignored or chastised god, no matter how many times you refused to believe in an after-life and no matter how you died - a nonillionth of a second after you die you will become one with your Soul and experience Heaven - in a timeless environment, for as long as you wish.

You won't worry if you've left loved ones behind, or if you've got unfinished business, or if someone said something really nasty to you before you died. You'll be in such a state of bliss that you won't care about that stuff. You couldn't even if you tried because you'll be in a GODlike state of constant joy. Worry, guilt and hurt are just not possible in this state.

Instead, you will look down at all your grieving relatives with GODlike understanding, compassion and a bemused smile as they wonder if you're OK, or stuck in Purgatory, or gone to Hell. They might even pay 'Wanda the Magnificent Psychic' a sum of money to contact your dead spirit and command your non-conscious chunk of subconscious mind to 'Go to the Light'. But you're *already there* as Light.

REINCARNATION

The entire death journey so far has only taken one nonillionth of a second - one pulse cycle. It's 'down, up, dead'. Only a nonillionth of a second has passed *since you passed*.

Your first task as a Holy Soul is to plan your next incarnation. You call your Soul-Group together for a meeting and discuss evolution, potential and challenge. You re-evaluate your recent life as well as past and future lives and access its accumulated potential and flow-on effect. You ponder your next move - what sort of challenge to take next, in order to re-member a predetermined level of potential and in accordance with the highest will and good of all Souls.

You might choose to 'go again' and reincarnate as a human-being with new parents and a new environment for a newer, greater understanding of joy. There is no space and time so you might go back to the same life, same body, same mindset as a 'walk in' to repossess your mindset when you were thirty-three years of age. How would *we* know?

You might decide to stay in the spirit-world for a while and guide other humans as a Soul. You might decide to join the ranks of Gods who co-create universes. You might decide to come back as a great spiritual leader, a spiritually aware public figure or even more powerful and amazing - a loving mother.

You might decide to come back as a rock and just contemplate the beauty of the desert floor for seventeen million years - in a state of bliss. You might decide to start again at the zero per cent end and work your way back from Neanderthal to normal over 25,630 years. You might decide to start again, as a Light being in a higher parallel dimension within an enlightened Cosmos.

Whatever you choose, the physical journey on Earth is always the same: co-create the extreme-duality of impurity, with a whole bunch of other humans, so that you have something to respond and react to. Then react badly, go downhill, feel bad, then challenge yourself to find the potential within yourself to cheer up again (get over yourself!).

Your responses and reactions to the extreme-duality you co-create, also help you find out what you like and don't like. This helps you know what you love so you can do more of what you love. Doing what you love because you know what you love doing then gives you the option of fast and enjoyable, en-Light-enment, rather than the slow and unnecessary pain and suffering of the EGO.

You do this lifetime after lifetime until you have re-membered enough potential to consciously respond to life with powerful and wise thought-action *every single time,* to maintain a constant state of joy, joyful service and love, as a human *being* Jesus-like. Only then will you transcend the need to reincarnate on Earth as a human.

Once you complete the Jesus-like 'Game of Life', you start a *new* 'Game' on a new level in a new dimension

(GOD knows where). And you keep doing that, lifetime after lifetime, incarnation after incarnation, dimension after dimension, until you reach perfection.

But here's the catch: you will NEVER reach perfection. Perfection is 100% purity of thought. That's GOD and that's already been done. As a Soul you exist in the Cosmos, which is not-GOD, which is *not-100%,* which is 99.999999999 ...%. That means the journey towards perfection is the journey towards 99.999999999 ...% and 99.999999999 ...% NEVER ends. It recurs for infinity. That means the journey towards perfection never ends. It recurs for infinity.

You will infinitely chase the recurring 99.999999999 ...% tail of purity as it beautifully and fractally cascades towards infinity, but you will never catch it. None of us ever will. The journey towards purity of thought will continue, onwards and upwards forever. It just gets better and better, each incarnation ... parabolically and *infinitely* as we head into higher and higher states of joy and love. Now that's yoooge!

YOU THEN MEET GOD

After you have chosen the challenge of the next incarnation, there is a spiritual party held in your honour, to celebrate your dualistic contribution to the expansion of the Cosmos, as a great spiritual warrior. The special guests of that party would include the First Being, the Gods, your Soul-Group and every other being of Light within the spirit-world. Now that's a party of Lights!

Suddenly the party will become still, perfectly still - a stillness beyond silence. Everything will become dark, timeless and peaceful. Every God will stand to attention, in awe and in a state of indescribable bliss because they know and you will know, that you are about to meet GOD.

But She won't come skipping down a marble staircase wearing a pink dress, throwing daises in the air chanting Om Namah Shivaya. Instead, She will *become you* and you will *become GOD*. In that moment you will experience all love from all beings from all time. Shortly after that you will realise that you *are* GOD.

SUMMARY

- En-joy life because it ends abruptly with death.
- The Cosmos is not out to kill you. You are in charge of your own e-motional suicide.
- The potential gained from the challenges of life are often more en-Light-ening than trying to unravel why you were breast-feeding a duck in your dreams.
- The first stage of death: your conscious mind enters the astral-world.
- The second stage of death: you re-evaluate your life in the dream world to find peace.
- You will stay in the astral-world until you find peace because you cannot continue onto the gateway of death until you find peace.
- Your Soul along with the prayers for the dying, spiritual work and the thoughts of grieving relatives will help you awaken from the nightmare of 'Hell' and lead you to the Cosmic-pond.
- The third stage of death: to enter the tranquillity and peace of the Cosmic-pond. This is the gateway of death and gives you the opportunity to either cross over into

the spirit-world to die or return to the physical-world to live.

- If you decide to cross over, then the fourth stage of death is to enter the spirit-world. Your pulse disintegrates, your heart stops and you are officially dead.
- Meanwhile your subconscious mind gets dropped off in the astral-world, because it's too impure to enter the spirit-world.
- This is the ghost.
- The ghost lies dormant until it's activated by a spare conscious mind, often the conscious mind of a grieving relative.
- Ghosts serve a valid purpose of helping grieving relatives interact with the memory of the dead person, to help them grieve.
- They can't interact with the real dead person because they become a Holy Soul and interacting with a Holy Soul would turn a human into a fried, bloody mist.
- A psychically sensitive psychic-medium can often help a grieving relative interact with a ghost. But it's not the real person - it's only their memory.
- The *really good* psychic-mediums *can* communicate on a Soul level, giving you 'direct' access to your Soul's downloads.
- Or - forget all that stuff: hand over the spirit-world needs to Soul and then concentrate all your attention on living here, more, now, in the physical world.
- True, personal, en-Light-enment is here on Earth, not in the spirit-world.

- The incredible amplitude within the e-motions of grief are often powerful enough to inspire enough energy into a ghost that it transcends dimensional separation and affects the real world.
- This is rare.
- More often than not, most 'evil' is caused by the intense amplitude of low-frequency vibrations, created by negative human e-motion.
- You will never know the exact time and place of your own death. Neither will anyone else. It's pre-sented to you, by your Soul-Group, as a surprise.
- Soul-Groups often use the shock value of death to help awaken humans to the joys of life.
- Only when you remove the fear of death will you start to live life because the fear of death is the major reason why people don't *challenge* themselves.
- Ask yourself, in any situation, 'What's the worst thing that can happen? The answer is death and that's awesome.
- The last stage of death is when your conscious mind reunites, once again, with your true subconscious mind - your Soul's memory-of-potential - to become wholly (Holy) *one* again.
- As a Holy Soul you will think like a God again, with 99.999999999 ...% potential and feel Cosmic joy and love. You will experience Heaven.
- You will then plan your next reincarnation.
- There will be one Heaven of a party.
- You will then meet GOD.

CONCLUSION

I'm sure it's more complicated than that but who cares!

GLOSSARY

10,000 Things - See 'The 10,000 Things'

Astral-World - A special layer of spirit-world that can tolerate the impurity of human thought. It could be considered the lowest vibration of the spirit-world and it acts like a bridge between the physical-world and the spirit-world.

CAPABILITY - The potential-aspect of power that influences the brain to think:

- I am an amazing and talented person who is capable of anything and worthy of everything. I have the strength, willpower and skills to achieve my goals to do, be, have whatever I desire.
- I choose people in my life that help, guide and support me, to be what I choose to be. I am appreciated by everyone in every way.
- The world supports my success.
- Everything I need to achieve my goals will be supplied by the unlimited creative power of my Soul.
- I AM CAPABLE.

Conscious-Mind - The role of the conscious mind is to respond to life with security and capability. The potential-aspects of security and capability are contained within the conscious mind of power and influence the brain to consciously think security and capability thoughts, *now*. Consequently the conscious mind is the power-mind, the mind of *now*, or the *power of now.*

Contemplation-Mindset - Your Soul thinks the *thought-of-the-human-being.* That thought comes alive (ONE) and contemplates its own power and wisdom. The contemplation-mindset pulses back-and-forth between the physical-world and the spirit-world to co-create the vibrational-reality of the human-being that extends from the human body to the aura to the astral-world to the spirit-world. The contemplation-mindset always contemplates itself in the background and that contemplation will always create the vibrational-reality of the human-being, even long after you have died (your dead body, the after-death aura, the vibrational-reality of the rotting body, your legacy, etc).

Cosmos - The original 'I am that I am' thought of GOD that came alive and thought for itself as the first, separate conscious living being. Its ONE-TWO-THREE-10,000-things contemplation distorted nothingness, adding space-time to nothingness, transforming that nothingness into ripples of space-time.

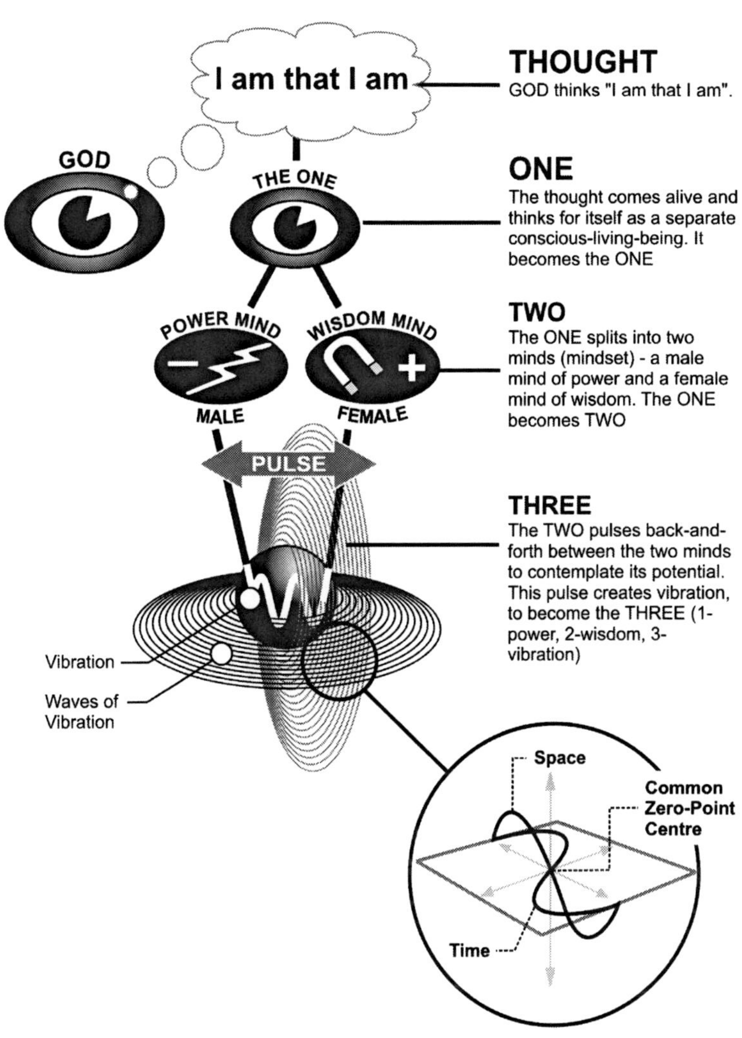

Cosmic-pond - A flexible, organic, grid of space-time; created by the Cosmos's ONE-TWO-THREE-10,000-things contemplation; that supports the reality of the physical-world, the astral-world and the spirit-world. The grid organises electromagnetic vibration into form, function and phenomena - a living 'manifestation of 'I am that I am' that's self-aware, has mind-blowing levels of potential and thinks for itself. The Cosmic-pond decides what form, function and phenomenon is required, depending on the interacting thoughts of other conscious-living-beings.

Divine Magnificence - The Divine Magnificence of GOD expressed as 100% pure potential.

Doing what you love because you know what you love doing - Purposefully challenging yourself with an enjoyable hobby, career, purpose, calling or dream; and doing so because you're secure and capable enough *to do* what you love; because your expressive and knowing enough *to know* what you love.

Download - A download is like an Internet download, where a lot of information arrives *all at once, within* each person's mind via their Soul, often experienced as a block of thought accompanied by a feeling to create a knowing. These are the 'Wow' moments where you just know stuff, it's mind-blowing, it feels amazing and you have to go, 'Wow!'

Dream-World - Where your conscious mind goes when you dream, same place as the astral-world.

Duality - For every experience there is an equal and opposite other-experience that helps us recognise and appreciate what the first experience was, by comparing the contrast of the two experiences together. Otherwise, how would we know? See also extreme-duality.

EGO - The EGO is a bad, subconscious habit of reacting to life with negative thought. It makes adults subconsciously behave like a poorly behave child (fear, failure, depression and ignorance, seriousness and hate).

E-motion - The ripples in the Cosmic-pond created by thought-waves, or potential energy-in-motion, or e-motion for short. E-motion floods the body with electromagnetic ripples, everywhere-all-at-once, which we experience as the real, physical sensation of feeling.

En-Light-enment - The slow re-membering of potential that adds more Light (re-membered powerful and wise potential) to the mind so you become en-Light-ened

EXPRESSION - The potential-aspect of wisdom that influences the brain to think:

- This is me. I am unique, I am important and I matter. I easily express myself, my emotions and my point of view.
- My ideas, point of view and boundaries are valid and will be respected by all. I speak up to create and embrace change in my life. I say 'Yes' and I can say 'No'.
- I share my inspiration and desires with the world.
- I am artistic, creative and intelligent.
- I EXPRESS MYSELF.

Electrical-Potential - Primarily an electrical force. Electrical-potential urges the human brain to think thoughts of security (survival, courage and abundance) and capability (skill, success and willpower) and gives a male characteristic to mindset.

Extreme-Duality - For every experience there is an equal and opposite other-experience that helps us recognise and appreciate what the first experience was, by comparing the contrast of the two experiences together. This is called duality. The greater the duality, the greater the contrast, the greater the experience. This is called extreme-duality.

Forgetting the Truth - As a human you've forgotten you're a God, so you believe you are a human-being. You've forgotten you are one with all things, so you believe you are separate from all things. You've forgotten the Cosmos is an illusion, so you believe reality is real. The combination of the three creates the unequivocal belief that you are a separate, human-being, living within reality. This is forgetting the Truth (with a capital 'T' since it's a Divine Truth). The Truth is that you are a God with amnesia, connected to all living things, participating in a giant illusion.

General Unhappiness - General unhappiness is when people think they're happy but really their emotional roller-coaster is mostly bouncing off the lower limits of happiness keeping them trapped in a general state of unhappiness. They've become so used to this emotional state that they consider it to be 'normal' and so assume they're happy. But it's not 'normal' and they're not happy. It's general-unhappiness created by the EGO taking over, due to a lack of potential challenge.

Ghost - After you die, your subconscious mind lies dormant in the astral-world. It's dormant because it no longer has a conscious mind and so can no longer pulse as a mindset. A non-pulsing mindset is a non-conscious being. Without its conscious mind-partner, it's just the memory of the person - a non-conscious, non-pulsing, disconnected chunk of subconscious mind, which I call a ghost.

GOD - That which *is* the real GOD. GODlike refers to anything that is more *like* the true concept of GOD and not-GOD refers to that which is not like the true concept of GOD. It's also the *only* Divine element that gets a capital personalisation such as GOD but as well as 'Her', 'She', 'I', 'It', etc.

god - That which people *think* is the real GOD, commonly known as 'God', the dude with a beard sitting on a cloud that's described in the Bible. It gets a lower case 'g' because *that* idea of god is essentially flawed and so does not deserve a capital 'g' - in my book.

God - God refers to the mythical Gods, the more balanced concept of God/Goddess, the Soul (Higher-Self) and Soul-Group. They all get a capital 'G' because those concepts are much closer to what *should* be worshipped as a deity(s). As do other GODlike elements such as Mother Earth, the Moon and the Sun

GODlike - GODlike refers to anything that is more like the true concept of GOD, or to be like GOD, or as close to GOD as you can be within an illusion of not-GOD (0-99% GODlike). In human terms, being GODlike is thinking-doing-feeling with power and wisdom so that you feel joy and love *most of the time*

Human-Mindset - The same potential within the contemplation-mindset, urges the human brain to think powerful and wise thoughts on a human level (human-mindset) and at a level that matches the potential you have re-membered (memory-of-potential). The human-mindset pulses back-and-forth between the physical-world and the astral-world and that pulse produces a vibrational-reality that manifests as events in your *you*niverse.

Illusion - To solve the dilemma of GOD not being able to fully experience the Divine Magnificence of Herself, GOD decided to create something that was not GOD, so she could have something to compare Herself to. You cannot have not-GOD is GOD is everything, everywhere and everywhen, so GOD created the *illusion* of not-GOD.

Impurity - Impurity of thought obtained by subconsciously reacting to life with bad habits of negative thought, producing weak and foolish thought-action to feel seriousness and hate - negativity.

KNOWING - The potential-aspect that influences the brain to think:

- I know that I am secure, capable, expressive and knowing.
- I know how to have fantastic, trusting, loving relationships with people.
- I know my life's purpose and my higher purpose. I know the next step. I trust my intuition and make perfect decisions every time. I know I am guided by my Soul.
- I know what I like and what I don't like. I know what to do, be and have.
- I KNOW.

Light - Light (capital 'L') is electromagnetic vibration, or potential in motion, or thought-waves, or e-motion. Light makes up all reality and is what I saw in November 2003.

It's also the Light (electromagnetic vibration) I saw in November 2003!

Magnetic-Potential - Primarily a magnetic force. Magnetic-potential urges the human brain to think thoughts of expression (personality, individuality, fulfilment) and knowing (creativity, imagination and life-experience) and gives a female characteristic to the mindset.

Memory-of-Potential - Your Soul is a giant subconscious mind. It acts like a spiritual diary and keeps a cumulative record of the potential you have re-membered from each challenge. It stores the powerful and wise potential (the gold) *behind* every thought from every nonillionth moment of every past, present and future challenge - including your wishes, imagination and dreams - as a memory-of-potential

Mind - A container that holds potential. There are two minds: a mind of power and a mind of wisdom. The power-mind holds the potential of power. The wisdom-mind holds the potential of wisdom.

Mindset - The two minds (containers) of power and wisdom combined. One mind contains the potentials of power and the other mind contains the potential of wisdom.

Necessary-Challenge - To ensure an appropriate level of en-Light-enment, if you don't voluntarily challenge yourself purposefully, your Soul will challenge you, *for* you, in a necessary way - a necessary challenge. This is the 'forced' path of en-Light-enment and it often involves unnecessary pain and suffering (essential suffering).

Nonillion - Nonillion is 10 to the power of 30, or a 1 with 30 zeros behind it.

Nothingness - What GOD was (and is) before Creation. 100% pure potential, existing as still nothingness, without vibration. Nothingness is the GOD that exists within all things.

Not-GOD - Not-GOD mostly refers to anything that is not GOD, which is anything that is not 100% pure potential, which could be anything that is 0-99% GODlike. It can also refer to that which is not like the true concept of GOD.

ONE - See 'The One'

OPTA - OPTA stands for *only positive thought action* and is a good habit of responding to life by consciously choosing positive thought. OPTA makes adults respond with security, capability, expression, knowing, joy and love.

Parabolic Change - A change that starts out slow and gradually increases to a crescendo, to form a curve on a graph, rather than a straight line.

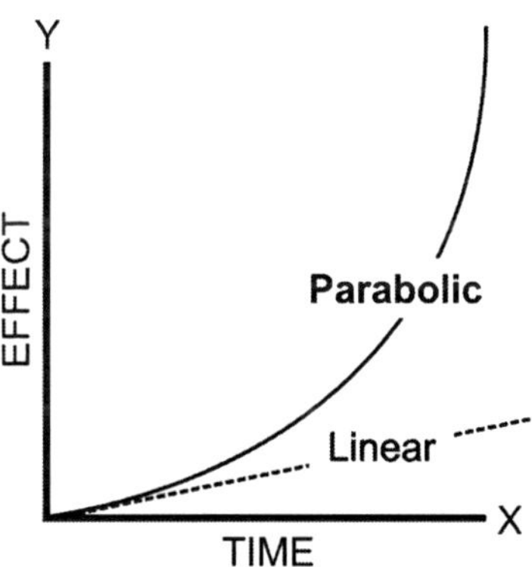

Physical-World - The physical-world is the real world we live in and includes what we can see and touch and interact with such as objects, biology, sunlight and gravity.

Potential - Potential is an electromagnetic *force* that urges thought. The human brain thinks, after it is urged by the force of potential. There are two main potentials - wisdom and power. These potentials break down further into the potential-aspects of security, capability, expression and knowing.

Potential-Age - A new era of human evolution where the entire human race will collectively embrace more power and wisdom; and so think and act with more security, capability, expression and knowing; producing a parabolic increase in joy and love across the planet. The potential-aspect begins on Dec 21, 2012.

Potential in Motion - Potential in motion is what happens when the potentials *behind* thought are converted into vibration, via the pulse of contemplation (thinking).

POWER - Describes both the mind of power and the potential of power contained *within* that mind that urges the brain to think thoughts of security (survival, courage and abundance) and capability (skill, success and willpower), Power gives a male characteristic to the mindset.

Power-Mind (Mind of Power) - Part of the mindset that contains the potentials of power, made up of the two potential-aspects of security and capability. The potential within the power-mind influences the human brain to think with security and capability.

Purity - Purity of thought obtained by consciously choosing to respond to life with powerful and wise thought-action to feel joy and love - positivity.

Purposeful Challenge - Voluntarily challenging your self in a demanding, exciting and enjoyable way such as a hobby, career, parenting or vocation.

Respond - Consciously responding to life by consciously choosing what to think, such as choices, decisions, skills and willpower. Good responses from the basis of positivity.

React - Subconsciously reacting to life using automatic habits of thought, stored in the subconscious memory (reacting without thinking consciously thinking about it) such as habit, talent, creativity and personality. Bad reactions form the basis of negativity.

Re-member - To re-member the potential that YOU chose to forget - 're-membering' as in putting things back together, the opposite of 'dismember'. You re-member potential by facilitating, mastering or conquering a challenge (purposeful-challenge or necessary-challenge). The more potential you re-member, the more powerful and wise potential within you r mind, the more Light within the mind, the more en-Light-ened you are. The purpose of life is to re-member your forgotten potential.

SECURITY - The potential-aspect of power that influences the brain to think:

- I am brave and strong.
- I trust people; they are supportive and kind.
- My world is a safe and wonderful place.
- All my wants and needs are supplied by the unlimited creative power of my Soul. There is no limit. I can do, be, or have whatever I desire.
- I AM SECURE.

Sex-mental-attraction - The primal urge of our conscious mind to reunite with its 'missing' female-subconscious mind partner (usually an external, surrogate, subconscious mind of *another* person)

Soul - Your Higher-Self or God-Self that lives in the spirit-world. This is the real YOU.

Space-time - Space and time work in unison, one cannot exist without the other, hence the term space-time. The distortion of space moves up-and-down over nothingness. The distortion of time moves side-to-side over nothingness. Both distortions share the same zero-point centre and that common zero-point re-combines the two distortions together so they both *share* space and time.

Spirit-World - The spirit-world is the unreal world which includes the unseen high-vibrational realms of reality that we cannot see or interact with that accommodate your Soul, memory-of-potential and the Gods.

Subconscious-Mind - The role of the subconscious mind is to record every thought you have ever made. Those thoughts accumulate in the subconscious mind to become automatic habits of thought - memory. The potentials of expression and knowing stored in the subconscious mind help you to store memories, retrieve memories and then react to life, based on the memory of how you have previously responded to life. The subconscious mind is your automatic mind. It operates without our conscious consideration and takes care of all the 'repetitive' thought-actions, such as instinct, talent, speech, personality and imagination. Consequently, the subconscious mind is the wisdom-mind, the mind of *not-now* or the wisdom of *not-now*.

The ONE - The potential within a thought urges that thought to think for itself. A thinking being is a conscious, living being, so when a thought thinks, it comes alive as a self-aware, separate, conscious living being. It becomes the ONE.

The TWO - The first thing the ONE thinks about is its own powerful and wise potential. To do so, and avoid the potentials behind the contemplation cancelling out to nothingness, the ONE splits into two minds. It becomes the TWO.

The THREE - The TWO then pulses between its two minds, to separately embrace its two potentials, to think the powerful and wise thoughts necessary to contemplate its potential. The pulse of that contemplation creates ripples of electromagnetic vibration in the Cosmic-pond. The TWO becomes THREE (vibration).

The 10,000 Things - The ripples of the THREE interact with the ripples of all other beings and the ripples of space-time to co-create something, somewhere, somewhen. The THREE becomes 10,000 things.

THREE - See 'The THREE'

TWO - See 'The TWO'

Thought-Waves - The ripples produced in the Cosmic-pond as the potentials *behind* thought are transferred onto the Cosmic-pond by the pulse of thought. Thought-waves are potential in motion, or e-motion.

Truth - As a human you've forgotten you're a God, so you believe you are a human-being. You've forgotten you are one with all things, so you believe you are separate from all things. You've forgotten the Cosmos is an illusion, so you believe reality is real. The combination of the three creates the unequivocal belief that you are a separate, human-being, living within reality. This is forgetting the Truth (Truth with a capital 'T' since it's a Divine Truth). The Truth is that you are a God with amnesia, connected to all living things, participating in a giant illusion.

Vibration - Potential in motion, or thought-waves, or e-motion. It's what happens when potential moves because it's been expressed as thought. All reality is made of vibration, hence vibrational-reality. Vibration is magnetic-potential through space over electrical-potential through time or what scientists call a 'linearly self-propagating transverse oscillating wave of electric and magnetic fields' - electromagnetic vibration for short.

Vibrational Reality - The 10,000 things reality of the physical-world, astral-world and the spirit-world, produced by ONE-TWO-THREE contemplation of conscious-living-beings. The Gods produce the vibrational-reality of the Cosmos. Humans produce the vibrational-reality of phenomena - events in the *you*niverse.

YOU - Your Soul or Higher Self

***You*niverse** - Our thoughts co-create events in our world and we are the centre of that world, or *you*niverse. Our own individual vibrational-reality is co-created when the ripples of our thoughts interact with the nonillions of other ripples from the thought-waves of millions of people thinking thousands of thoughts; to co-create by nonillions of points of double amplitude - eventual reality. That reality can manifest into our world as any sort of event such as luck, support, opportunities, ideas, accidents or even violence. The vibrational-reality is more real to the thinker because the amplitude of the thought is strongest at the source. In that regard, we are all the centre of our own private universe, or youniverse.

WISDOM - Describes both the mind of wisdom and the potential of wisdom contained *within* that mind that urges the brain to think thoughts of expression (personality, individuality, fulfilment) and knowing (creativity, imagination and life-experience). Wisdom gives a female characteristic to the mindset.

Wisdom-Mind (mind of wisdom) - Part of the mindset that contains the potentials of wisdom, made up of the two potential-aspects of expression and knowing. The potential within the wisdom-mind influences the human brain to think with expression and knowing.

READING LISTS

These are the recommended reading lists I mentioned earlier. 'The Books that Changed My Life' reading list contains books and authors mentioned in the 'Acknowledgements'. As the title says, these books really changed my life, when I first started on this journey. These books are listed in alphabetical order, although the books that *really* changed my life in a dramatic way are in **bold.**

The 'Metaphysics of Dis-ease' reading list contains fantastic information about the metaphysics of dis-ease, written by people more qualified in this field than myself. I have listed them in order of my personal preference. If you only get time to read one book, then I highly recommend, 'Sanctuary: The Path to Consciousness' by Stephen Lewis and Evan Slawson'.

BOOKS THAT CHANGED MY LIFE

- A New Earth: Awakening to Your Life's Purpose by Eckhart Tolle.
- Anastasia - The Ringing Cedars of Russia Series (9 volume set) by Vladimir Megre.
- Anatomy of the Spirit by Caroline Myss Ph.D.
- **Ask & It Is Given by Esther and Jerry Hicks.**
- **Conversations with God: An Uncommon Dialogue (Books 1, 2 & 3) by Neale Donald Walsch.**
- Crossing Over: The Stories Behind the Stories by John Edward.

- **Do What You Love, The Money Will Follow: Discovering Your Right Livelihood by Marsha Sinetar.**
- Feel the Fear and Do It Anyway by Susan Jeffers.
- Hands of Light - A Guide to Healing Through the Human Energy Field by Barbara Ann Brennan.
- Heal Your Body A-Z: The Mental Causes for Physical Illness and the Way to Overcome Them by Louise L. Hay.
- Heaven and Earth: Making the Psychic Connection by James Van Praagh.
- Love is in the Earth: A Kaleidoscope of Crystals by Melody.
- Messages from Your Angels by Doreen Virtue Ph.D.
- Opening to Channel: How to Connect with Your Guide by Sanaya Roman and Duane Packer.
- Sacred Contracts: Awakening Your Divine Potential by Caroline Myss.
- **Sanctuary: The Path to Consciousness by Stephen Lewis and Evan Slawson.**
- **The Amazing Power of Deliberate Intent: Living the Art of Allowing by Esther and Jerry Hicks.**
- **The Astonishing Power of Emotions by Esther and Jerry Hicks.**
- **The Body is the Barometer of the Soul, So Be Your Own Doctor by Annette Noontil.**
- **The Celestine Prophecy by James Redfield.**
- The Crystal Bible by Judy Hall.
- The Four Agreements by Don Miguel Ruiz
- The 4-Hour Workweek: Escape 9-5, Live Anywhere, and Join the New Rich by Timothy Ferriss.

- The Hidden Messages in Water by Masaru Emoto and David A Thayne.
- **The Law of Attraction: The Basics of the Teachings of Abraham by Esther and Jerry Hicks.**
- The Lightworker's Way: Awakening Your Spiritual Power to Know and Heal by Doreen Virtue Ph.D.
- The Pleiadian Agenda: A New Cosmology for the Age of Light by Barbara Hand Clow and Brian Swimme.
- **The Power of Now: A Guide to Spiritual Enlightenment by Eckhart Tolle.**
- **The Road Less Travelled: A New Psychology of Love, Traditional Values and Spiritual Growth by M. Scott Peck.**
- The Reconnection: Heal Other's, Heal Yourself by Dr Eric Pearl.
- The Science of Getting Rich by Wallace D. Wattles.
- **The Secret by Rhonda Byrne.**
- The Seven Spiritual Laws of Success: A Practical Guide to the Fulfillment of Your Dreams by Deepak Chopra.
- The Starseed Transmissions by Ken Carey.
- The Tenth Insight: Holding the Vision (Celestine Prophecy) by James Redfield.
- Think and Grow Rich by Napoleon Hill.
- Way of the Peaceful Warrior: A Book That Changes Lives by Dan Millman.
- **What the Bleep Do We Know!?: Discovering the Endless Possibilities for Altering Your Everyday Reality by William Arntz, Betsy Chasse and Mark Vicente.**
- You Can Heal Your Life by Louise L. Hay.

THE METAPHYSICS OF DIS-EASE

1. **Sanctuary: The Path to Consciousness by Stephen Lewis and Evan Slawson.**
2. **The Body is the Barometer of the Soul, So Be Your Own Doctor by Annette Noontil.**
3. Heal Your Body A-Z: The Mental Causes for Physical Illness and the Way to Overcome Them By Louise L. Hay.
4. The Reconnection: Heal Others, Heal Yourself by Dr. Eric Pearl.
5. Anatomy of the Spirit by Caroline Myss, Ph.D.

All the above titles in the reading lists are available at our sister website ITSSOULED, along with all the popular spiritual, new age, alternative and holistic books, DVD and CDs. We only list products at ITSSOULED that have been deemed amazing, life-changing, inspiring or enlightening.

ITSSOULED

www.itssouled.com.

INDEX

ABOUT YOOOGE

Yooogе was founded by Matthew John Corcoran after a yooogе experience in November 2003 changed his life forever. Yooogе is a family-operated, independent, non-religious organisation, dedicated to the teaching of joy. The aim of Yooogе is to help people *feel good* - physically, emotionally and mentally. Yooogе also teaches the art of irresponsibility and childish silliness via its range of funny t-shirts and stickers. For more information visit the Yooogе website at www.yoooge.com.

ABOUT THE AUTHOR

Matthew John Corcoran was born in Perth, Western Australia. He has worked as a designer, stand-up comedian, professional musician and is now a writer, public speaker and spiritual entertainer. He has a Masters Degree in Theoretical Metaphysics from the 'University of Potential Life Experience' that he awarded to himself when he turned forty-two years of age. He believes that everyone should have one ... and everyone can have one if they print one out for themselves. He lives and works in coastal New South Wales, Australia, on the beach mostly (living the dream!) with his partner Jayne.

THE YOOOGE MEMBERS SECTION

Become a member of Yooogе. It is free to join and members get the Yooogе newsletter - an irregular e-newsletter than keeps members up to date with Yoooge goodies such as downloads of new information, news, events and freebies. You can join by going to the Yoooge website at www.yoooge.com and clicking on the 'Members' button.

YOOOGE BOOK ONE

'The Yoooge Philosophy of Life' is a deeply spiritual, non-religious explanation of the purpose and meaning of life. It explains the real truth about GOD, the Gods and god and who really created the Cosmos and why. It explains the agreement we made with our Souls and our role as humans in the expansion of the Cosmos. It explains why there is so much suffering in the world and how we can transcend that suffering with power, wisdom and joy. It explains why we think the way we do and the real purpose of e-motion. It explains our life purpose, our other purpose and our higher purpose. It explains what's really going to happen in 2012, what happens when you die and what happens after you die. Available now as a hardback, paperback and eBook.

YOOOGE BOOK TWO

'The Yoooge Journey of Life' explains the development, characteristics and deep spiritual purpose of the EGO. It explains how the EGO is developed in childhood and how it follows us into adulthood as a bad, subconscious habit of reacting with negative thought. It explains how to recognise the EGO and how to resolve it. It reveals how the journey of life unfolds in cycles of seven years and how to find where you are in this journey. It reveals the mystery behind human nature and *why* we feel good, *why* we feel bad and how we can feel good *all the time.* Coming soon. Go to www.yoooge.com and join the 'Yoooge Members' section for more details.

YOOOGE BOOK THREE

'The Yoooge Experience of Life' details the astonishing co-creative power all humans have and how we can tap into this power to create yoooge experiences - events, opportunities, synchronicities or anything else you need to facilitate your purpose. It explains 'The Yoooge Pulse Theory' and the mind-blowing reality of observation. It explains the process of creation, the role we play and the role our Soul plays. It shows us *how to co-create* purposefully and consciously rather than accidentally and unconsciously in order to do/be/have whatever we desire. It's the handbook of miracles. Coming soon. Go to www.yoooge.com and join the 'Yoooge Members' section for more details.

THE YOOOGE WORKSHOP

'The Yoooge Workshop' teaches the practical application of the Yoooge Philosophy. If you consider 'The Yoooge Philosophy of life' to be the 'Why?', then 'The Yoooge Workshop' is the 'How?' It's explains *how* to embrace your power and wisdom; *how* to think and act with security, capability, expression and knowing; and *how* to maintain a constant state of joy.

The Yoooge Workshop is a collection of some of the most powerful personal growth tools and processes on the planet, specifically designed to help people conquer their fears, achieve their goals, express their truth, discover their purpose and find true happiness/joy. We also added extra tools and processes that teach people how to resolve the

EGO, practise OPTA, get 'downloads' and have better relationships. Visit www.yoooge.com for more details.

FRIENDS OF YOOOGE

We're always looking for spiritually-minded, joyful people to help us 'spread the word' about Yoooge. If you really liked this book and you want to help us sell and distribute copies of the book (we'll pay you) then visit www.yoooge.com and click on the 'Friends' link.

HOW CAN I GET ANOTHER COPY?

This book is distributed by 'Friends of Yoooge'. To order more copies of this book, or other great Yoooge products, please contact the 'Friend of Yoooge' that initially recommended this book, or visit www.yoooge.com for more details.

Recycle like your
life depended on it!

CPSIA information can be obtained at www.ICGtesting.com
Printed in the USA
LVOW121522030912

297185LV00001B/126/P